CONSTRUCTIVE ERGONOMICS

CONSTRUCTIVE ERGONOMICS

edited by Pierre Falzon

CRC Press
Taylor & Francis Group
Boca Raton London New York

CRC Press is an imprint of the
Taylor & Francis Group, an **informa** business

This book was previously published in French by Presses Universitaires de France.

CRC Press
Taylor & Francis Group
6000 Broken Sound Parkway NW, Suite 300
Boca Raton, FL 33487-2742

First issued in paperback 2019

© 2015 by Taylor & Francis Group, LLC
CRC Press is an imprint of Taylor & Francis Group, an Informa business

No claim to original U.S. Government works

ISBN-13: 978-1-4822-3562-3 (hbk)
ISBN-13: 978-0-367-37832-5 (pbk)

Library of Congress Cataloging-in-Publication Data

Constructive ergonomics / edited by Pierre Falzon.
       pages cm
    Includes bibliographical references and index.
    ISBN 978-1-4822-3562-3 (hardback)
    1. Human engineering. I. Falzon, Pierre.

T59.7.C6645 2014
620.8'2--dc23   2014027317

Visit the Taylor & Francis Web site at
http://www.taylorandfrancis.com

and the CRC Press Web site at
http://www.crcpress.com

Contents

Section I: Resources and conditions for development

Introduction—Constructive ergonomics: A manifesto

Pierre Falzon

Fitting work to the human?

Since its early beginnings, ergonomics has set its goal as fitting jobs, environments and machines to the human. The symposium that led to the creation of the International Ergonomics Association, which took place in 1957 in Leyden, the Netherlands, was thus called 'Fitting the Job to the Worker', and the title of one of the earliest ergonomics books in France could be translated as *Fitting Machine to Man* (Faverge et al., 1958). Today, this goal certainly remains commendable – but is it enough? Does it provide an adequate response to the needs of people, societies and organizations?

This book aims to provide new answers to these questions, starting with the following statement: ergonomics cannot remain content with a limited and static view of *adaptation*, a view that would restrict its goal to designing systems that are suited to work as it is defined at a certain point in time, to workers as they are at a particular moment, and to organizations as they operate here and now.

The objective of ergonomics must be *development*: the development of individuals, based on setting up situations or action that lead to increased success and to the acquisition or construction of know-how, knowledge and skills; and the development of organizations, based on integrating, within these very organizations, reflective processes that are open to the workers' own capacity for innovation. Fostering the development of individuals as well as the development of organizations is only possible if individuals have sufficient operational leeway and freedom of action. This freedom of action includes the ability to continuously build and rebuild the rules of work.

This book is not a handbook. Its goal is not to present a comprehensive overview of ergonomics, nor of its concepts, models or methods. It is a manifesto that aims to redefine the ambition of the discipline and to describe some of the elements of its scope.

It advocates a constructive and developmental view of ergonomics. By *constructive* and *developmental* – in this book, both terms will be used interchangeably – we mean to highlight the fact that individuals as well as collectives of operators develop by interacting with the world, and by acting upon it. Their actions aim both to understand the world and to transform it. It is the constructive and developmental activity of subjects that constitutes the driving force of learning, transformation and performance. In contrast to a defensive approach to ergonomics, which would view work mostly as a source of constraints, and the role of ergonomics as reducing these constraints, the goal of constructive ergonomics is to eliminate obstacles hindering success and development. Constructive ergonomics aims to maximize opportunities.

As we will see, the stake for ergonomists is to develop the enabling potential of organizations, so that they might contribute simultaneously and sustainably to improving the well-being of employees, to encouraging the development of skills and to improving performance. Any organization has a more or less promising enabling potential. However, this potential is often underused, unknown or unrecognized. In some cases, it may even be hindered by the organization itself. The goal here is not to create a new 'enabling' task that would complement existing tasks, but to organize existing work so that it will enable individuals and organizations to make some progress (Falzon and Mollo, 2009).

Development as a fact, a purpose and means

Development as a fact

Let us begin by considering development as a fact: during and because of professional practice, operators and collectives develop two kinds of skills. On the one hand, they develop knowledge, know-how and strategies related to the task itself. On the other hand, they develop knowledge about themselves: which activities they have more or less mastered, what is the maximum workload that can be undertaken safely, what is the comfort zone of their professional practice, what strategies they rely on to make use of themselves, what heuristics are available to make the best use of their own resources, etc. The goal of these skills is twofold, as they aim for both performance and well-being. They allow operators to better achieve their goals, and to do so more efficiently, while avoiding hazardous situations and protecting themselves.

Furthermore, over the course of time, operators undergo transformations. This is not just because of ageing; it is also because their career paths may or may not provide them with opportunities for development.

These effects of time may be beneficial or detrimental to various degrees, depending on the concrete conditions in which professional activity is carried out. First, these conditions influence the decline or preservation of people. Second, they may encourage – or conversely, hinder – the acquisition of skills allowing operators to cope with work situations (e.g. know-how related to caution and strategies aiming to conserve resources), as well as the construction of collective practices for preservation and performance. The challenge then becomes this: how can one design work organizations that leave some operational leeway and some room for the development of skills, practices and methodologies that encourage the expression or the emergence of knowledge and know-how?

Development as a purpose

Therefore, ergonomists cannot remain content with viewing operators in the here and now. They must take an interest in the conditions of development, and in career and life paths. Hence, development is a purpose of ergonomic interventions. The issue here is to contribute to designing environments that allow human activity to develop in all of its aspects – gestural, cognitive and social – while constantly aiming for the optimal compromise between the objectives of well-being and performance (Falzon and Mas, 2007).

The concept of 'enabling environment' has been put forth following this view. As a model, it makes it possible to integrate the various levels of ergonomic action (Falzon, 2005; Falzon and Mollo, 2009; Pavageau et al., 2007). This model was developed based on the works of A. Sen (2009), particularly on the idea of 'capabilities' that he proposes. A *capability* is defined as a set of operations that is truly accessible to an individual. A capability implies the availability of a capacity (i.e. a piece of knowledge, of know-how), but does not amount to that alone. It also implies a genuine possibility to use this capacity. Thus, the use of a capacity implies favourable conditions and the existence of conversion factors, in the sense that a capacity is converted to a genuine possibility. As an example, and drawing from Sen, the right to vote is not a guarantee for an effective capability of voting. The necessary conditions for a capability of voting include a sufficient level of education, an effective and fair dissemination of political information, an efficient organization of election processes and the right to vote.

According to Sen, the goal of public policy is the development of capabilities. Similarly, ergonomists aim to put operators in situations where they will be capable of action, by acting on the conditions in which their activity is to be deployed. Thus, an enabling environment can be understood following three different points of view. The first is preventive, the second is universal, and the third is developmental.

From the *preventive* point of view, an enabling environment is an environment that does not have detrimental effects on individuals and that preserves their future abilities for action. Here, we find a standard aspect of ergonomics interventions as they are carried out today: the goal is to detect and prevent hazards, to eliminate exposure to toxic substances, or alternately, to task requirements that might, in the long run, lead to deficiencies or detrimental psychological effects, etc.

From the *universal* point of view, an enabling environment is one that takes into account differences between individuals (anthropometric characteristics, as well as differences related to age, gender or culture) and that aims to compensate for individual deficiencies related to ageing, illness or disability. It is therefore an environment that prevents exclusion and unemployment.

From the *developmental* point of view, an enabling environment is an environment that allows individuals and collectives to

- Succeed, i.e. to apply their abilities in an effective and fruitful manner. It is not just an environment that does not hinder abilities, but one which makes people capable.
- Develop new know-how and new knowledge, broaden their opportunities for action, and strengthen their control over their tasks and the ways in which these are carried out – in other words, their autonomy. An enabling environment is an environment for continuous learning.

Development as a means

Finally, development is a means for ergonomic interventions. Project management and innovation rely on stakeholders to take a step back from their own work practices. This is a necessary aspect of designing future work. This can be aided by the use of simulations, confrontations of practices or training methods. Ergonomic action then becomes an opportunity to begin a process of development and learning, whether this process serves the design of organizations or that of artifacts. From this point of view, the objective is both to foster processes of development throughout the ergonomic intervention itself and to design work systems that will promote development themselves. Hence, development viewed as a means serves development viewed as a goal.

The latter point has one methodological consequence. Ergonomists cannot promote development as a goal of the discipline without advocating, in turn, methodologies of ergonomic intervention that encourage development themselves. Active involvement of operators in processes of organizational change and design is not an additional or optional feature of ergonomic interventions. It is a necessity, in order to ensure the consistency of a constructive approach.

This affects the professional position of ergonomists in major ways. Following a (very) traditional view of the field, the ergonomist is involved as an expert in human factors, based on his or her general knowledge about humans. The role of the ergonomist is to advise decision-makers (project managers, designers, managers). Following a more comprehensively equipped view of ergonomics, ergonomists add to this general knowledge about humans further knowledge derived from activity analysis. The ergonomist then becomes a representative of operators from the point of view of project stakeholders, and a designer amongst designers. In the view promoted here, the ergonomist becomes the linchpin of a participatory design process, which is itself developmental, and aims to achieve several goals at the same time: to transform the representations of all stakeholders – operators, managers, supervisors, staff representatives, etc. – and to achieve a satisfactory result – satisfactory implying here that the situation produced allows development to go on.

This position does not imply in any way that ergonomists should abandon the expertise that is their own. They should retain their knowledge about the effects of specific forms of work organization on human activity, about the methods that are useful for work analysis, about work in the situations that they have analyzed or about work in other similar situations, and knowledge about design and design methods. Depending on the need of the hour, the ergonomist will rely on this knowledge, when judged useful for the project to move forward. Therefore, this does not involve resigning in any way from the profession of ergonomics, where ergonomists would act as mere facilitators. On the contrary, it implies setting new targets, where the ergonomist's personal knowledge is placed in the service of a developmental approach.

Developing individuals, collectives and organizations

The central status of activity

Activity-centred ergonomics was built on a model that distinguishes a task from an activity. This model views an activity as the product of a continuous process that takes place within the subject. In this model, the operator is viewed not just as the person who carries out a task, but as the creator of his or her own mobilization. This mobilization relates, in a context-sensitive way, the requirements of the task to the subject's will for self-protection, success and learning. The operator regulates his or her activity depending on the results it produces, in terms of both achieving task goals and the effect this activity has on the operator and on collectives of operators (Falzon and Teiger, 1995; Wisner, 1995).

This model may seem very humdrum to some because it is so closely tied to the way in which they think of human work activity. And yet, it is very far from being universally shared within the international ergonomics community. This model, however, is operant; it allows us to understand human activity as well as its effects: both the effects that are detrimental – because they hinder or otherwise constrain regulation processes – and the effects that allow operators to be satisfied with their work and to make some progress. This is a model of an active subject, who is involved both in carrying out work and with protecting and transforming him or herself. It is opposite to the model of humans as passively carrying out a prescribed task.

In the past, the latter model has long prevailed, and indeed, it often still does. Although it has never allowed comprehending work activity in any relevant way, it is the basis upon which Taylorian or Neo-Taylorian organizations have envisioned work, even in their most recent incarnations. The increased demand of companies for greater quality (i.e. fewer defects) is still often dealt with by increasing prescription (e.g. Taylorization, total quality management, and more recently, Lean manufacturing processes and the development of evidence-based medicine) – in other words, by restricting operational leeway to a greater extent.

This view seems incapable of achieving alone the required levels of system performance. Every day, work analysts note the constant contribution of operators to adjusting work situations and adapting rules. This contribution is often seen in a negative way, as a violation of prescribed rules. Instead, it should be viewed positively and encouraged. Autonomy in decision-making, adaptive quality and safety are all required in order to cope with variability, to optimize processes, and to assist in the completion of work goals.

A constructive model of activity

Here, we return to a classical model of the regulation of human activity (see Figure I.1), originally proposed by Leplat (1977), in order to adapt this model to the constructive view we aim to defend here. Let us first remind the reader of its main characteristics:

1. The model distinguishes task from activity. A task is defined permanently by a set of goals, a level of requirement, means, criteria that need to be observed, etc., and temporarily by specific instructions, by the workload at any point in time, etc. An activity refers to a mobilization of the subject. This mobilization relates only indirectly to the task itself. The operator couples the prescribed task with

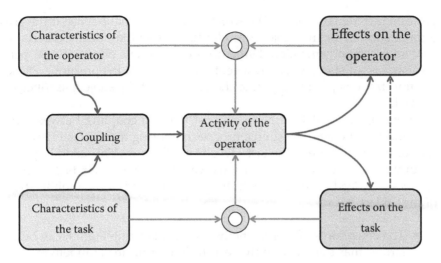

Figure I.1 The model of regulation of human activity. (Translated and adapted from Leplat, J., *L'analyse psychologique de l'activité ergonomie*, Octarès, Toulouse, France, 2000.)

elements that are specifically his or her own features: skills, representations of the profession, state of being at a certain point in time. From this coupling arise both the actual task – i.e. that which the operator defines for himself or herself – and the mobilization of the operator to complete this actual task.

2. The model distinguishes two types of effects of the activity: effects on the task (e.g. the level of task completion) and effects on the operator (e.g. fatigue).

3. The model proposes two feedback loops to characterize the regulation of activity. The first loop compares the operator's initial state to the state resulting from the operator's mobilization. The second loop compares the obtained results with the expected results. Once again, activity can be adjusted depending on the results of this comparison.

This model calls for several comments.

On the one hand, it should not make a clear-cut separation between the effects on the task and the effects on the subject. Success leads to satisfaction, and conversely, failure leads to frustration, hence the vertical arrow we have added connecting the effects on performance to the effects on the subject. It is worth noting that this connection is not mentioned as such in much of the literature in ergonomics. Mostly, performance tends to be viewed as a benefit for the system alone, as if the fact of succeeding did not lead to well-being. However, the question of the criteria of performance remains open. The words above only make sense if the criteria of success for the subject are identical, or similar, to the criteria of success for the organization. Difficulties or disorders emerge when performance is satisfactory with respect to the criteria of the prescribed task, but unsatisfactory with respect to the criteria that the subject ascribes to the task.

Second, the model seems to suggest that the coupling between the task and the operator is a simple process of pairing the characteristics of the subject with those of the task. This is not the case. This coupling may occur with some difficulty for various reasons. In particular, task characteristics may lead to difficulties in this coupling and in the mobilization of the subject:

- The constraints of work, whether material or immaterial, can be so strong that they leave only very little margin for manoeuvre. The coupling will then take place with little or no autonomy, and the possibilities for the regulation of activity will be almost nonexistent. Because of this, the operator's activity will be restricted to a single, repetitive way of doing things, with some well-known consequences on physical and psychological health: musculoskeletal disorders, occupational wear and tear and job dissatisfaction.
- Prescribed requirements can be at odds with the operator's own wishes. This conflict can prove insoluble, leading to conflicts in the subject's mobilization. The operator can be mobilized in spite of him or herself in order to achieve goals that he or she does not subscribe to. He or she can be led to carry out work that is at odds with his or her own standards of quality or personal ethics. From this point of view, psychosocial disorders can be viewed as pathologies of coupling. An impossible coupling leads to mobilizing the operator against him or herself, and to making development impossible.
- These difficulties become even more severe when these disparities and contradictions cannot be collectively debated within the organization. Each person then becomes locked in a solitary confrontation with the organization and with its prescriptions. These difficulties can only be resolved by an explicit debate concerning the rules of work involving both organizers and employees – in other words, through what de Terssac and Lompré (1996) term *cold regulation*.

Finally, and this is the most important point to us, the model aims to account for the subject's short-term functional activity. It features a subject that is unstable in the short run (e.g. because of fatigue, or in reaction to random events) but stable in the long run. Of course, this is insufficient, particularly considering the vision we wish to propose here.

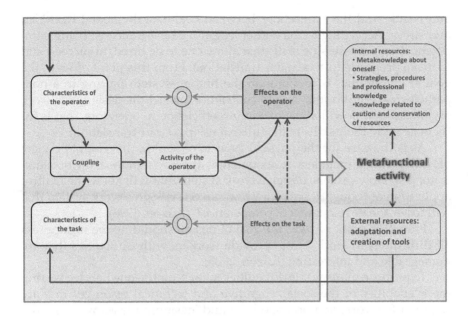

Figure I.2 Long-term regulation of activity.

Operators change because of ageing, but also because work alters them and because they alter work. These transformations may be detrimental – e.g. occupational diseases and accidents – but can also be beneficial – e.g. learning and development of new skills.

These beneficial transformations are the result of another regulatory process, which operates in the long term, and which Figure I.2 aims to illustrate. As subjects observe the effects of their activity on themselves, as well as the effectiveness or ineffectiveness of their strategies and their related costs, they elaborate internal resources: knowledge about oneself, new procedures and strategies – and external resources, and tools to assist work, either adapted from existing tools or created ex nihilo. Functional activity fuels metafunctional activity, which transforms the subject (Falzon, 1994).

Following the approach that we aim to advocate here, this second regulatory loop is crucial. The goal is to encourage metafunctional activities and the development of skills as much as possible. The latter is viewed as a necessity for both individuals and organizations.

Health, performance and development

The previous sections have established some connections between health, performance and development. M. de Montmollin (1993) set a milestone on this topic when he wrote one of the few existing texts concerning cognitive health. In this text, he examines the relationship between cognition and health, and the connections between health, skill, mental workload and stress. According to the author, cognitive health means 'being competent, that is, possessing skills that allow one to be hired, to succeed and to make progress' (p. xxxix, our translation). From this point of view, the goal of ergonomics is to maintain the human-system pairing in a non-pathological equilibrium, and to contribute toward the design of 'a work organization that will allow maximum efficiency to operators, that is for them to apply their skills to the full extent' (p. xl, our translation).

As is the case for the physical aspects of health, this cognitive view of health should include a developmental approach. Indeed, the question is not just 'How can one design a work system that will allow the fruitful application of thought?' but also 'How can one design a work system that will foster the development of competence?' (Falzon, 1996).

From this point of view, the goal of design should not be to suppress all difficulties in work, but to provide workers with difficulties that are manageable and interesting (Falzon, 2005).

To propose manageable difficulties means, on the one hand, ensuring the availability of the social, cognitive and technical resources that are required for work, and on the other hand, designing tasks with an adequate level of difficulty. Specific situations may prove to be unacceptable. This is related to an imbalance between resources and requirements, i.e. tasks with high requirements being carried out with insufficient resources.

To propose interesting difficulties means that operators will have to face situations with high stakes and overcome their difficulties, while developing new knowledge and know-how. Of course, this does not imply that every difficulty is an interesting difficulty. Operators often have to deal with uninteresting problems: ill-suited prescriptions, ineffective or incomprehensible procedures, unusable interfaces, impractical tools, etc. These difficulties are pointless and counterproductive in terms of both health and performance, and should be eliminated.

Human beings have a natural appetite for acquiring new skills. This is a constant observation for every ergonomist. Operators develop know-how, procedures and techniques because of and throughout the course of their work. This spontaneous inclination toward learning and discovery should be encouraged. It contributes both to the quality of the operator's work and to progress in organizations.

However, this appetite for learning has often been ignored by organizations. Moreover, these have made only timid attempts to organize work so as to encourage learning, although the concepts of human capital (Becker, 1964) and learning organization (Argyris and Schön, 1978) have contributed to making ideas move forward.

Switching from a vision of work grounded in qualification to that of work grounded in skill has been the first breach: what is expected is no longer to be able to abide by a prescription, but also to be able to react in relevant and self-sufficient ways to random and unpredictable events. A second related breach consists of a rising trend in prescribing work in terms of assignments, rather than in terms of tasks that need to be carried out. This evolution confirms changing expectations: operators are expected to define what they are to do. Both breaches have the same consequence: the sustained acquisition of new skills has become a condition of performance.

Therefore, acquiring skills has become crucial. Yet, skills are no longer constructed through repeated confrontations with identical or similar situations, but by confronting and analyzing singular situations. There has been a shift from implicit, only faintly conscious learning based on repetitiveness, to explicit, conscious learning grounded in reflective practices, which can be either individual or collective.

A guide to the development of the reader

The contributions gathered in this book present a strong overall consistency. They could have been organized in various ways. We have chosen to divide the book into two sections, based on the distinctions introduced previously on development as a fact, purpose and means. The first section, Resources and Conditions for Development, focuses on development as a fact and as a purpose, and examines the conditions that can encourage development. The second section, Dynamics of Action and Dynamics of Development, sets development as a purpose and as a means of action. Methodological developments have an important place in this second part.

Rather than present each chapter of the book in succession, we have proposed here a reader's guide following three major topics. Each of the chapters in the book can illustrate one or more of these topics.

Development as a factor of health and performance

The issue of skills and the conditions of their development is of crucial importance in this book. Various chapters focus on this issue and on conditions that are favourable or detrimental to the development of skills, which are viewed as driving factors of health and performance.

Catherine Delgoulet and Christine Vidal-Gomel revisit the fundamental distinction between productive activity and constructive activity, which corresponds to the distinction, introduced above, between functional and metafunctional activities. They identify some favourable conditions – conditions that may potentially foster development – and some unfavourable conditions for development, which combine a high degree of prescription with a high level of uncertainty.

Yannick Lémonie and Karine Chassaing address the development of professional gesture. They successively examine the mechanisms involved in the production of motor acts, the active part played by operators in the construction of gestures and gestural variability, viewed as a resource, and the conditions – which are reflective, and often collective – allowing these developments to take place.

Although the development of gestural and cognitive skills is a classical topic in ergonomics, this is not so much the case for the development of psychosocial resources. Yet, according to Laurent Van Belleghem, Sandro De Gasparo and Irène Gaillard, these resources should be seen as a component of the functional activity of subjects and of the mobilization of oneself, which is a requirement for any kind of work, just as one can speak of physical or cognitive mobilization. Hampering the development of psychosocial resources is no less pathogenic than hampering the development of gestures.

The collective aspects of development are highlighted on multiple occasions.

On the one hand, resources cannot be limited to only individual skills: the work collective is, in itself, a resource that is constructed over time. Beginning with the acknowledgement that the mere existence of a team is not enough for it to be considered as a collective, Sandrine Caroly and Flore Barcellini examine the conditions for the development of collectives and collective activity. This leads them to distinguish collective work from collective activity and work collectives. Justine Arnoud and Pierre Falzon focus on the conditions encouraging the setup of transverse collectives, i.e. collectives that involve different professions.

Furthermore, collectives contribute to the construction of resources. Thus, when learning a gesture, collectives make it possible to pass down a shared experience of the trade, as well as to conceptualize and put up for debate the gestures involved in work (Lémonie and Chassaing; Six-Touchard and Falzon). This is particularly useful in the case of tacit gestural know-hows, which cannot easily be put into words. The collective is also involved in the construction of skills to cope with night shift work. Every person knows the detrimental effects of night shift work and work in split shifts. Yet under specific conditions, operators may develop the know-how and skills that will allow them to protect themselves and succeed. According to Cathy Toupin, Béatrice Barthe and Sophie Prunier-Poulmaire, working time, although it can be constrained by scheduling, can also become a constructed time, provided that the work organization is open to transfers of practices within the work collective.

The issue of time is tackled in a different way in the chapter dedicated to career paths. Corinne Gaudart and Élise Ledoux focus on 'long' time, i.e. the kind of time involved in courses of work over the duration of a career. They argue in favour of a longitudinal analysis of these courses of work. Activity at any given time expresses the positive and negative experiences of the past and directs future development. Processes of decline and growth combine in a mutual dependency. Experience is the product of this constant reconstitution. Hence, passing on experience is a developmental task.

Developing organizations through the development of practices

The issue of organizations intersects with several chapters, if not most of them. Two points appear to be of central importance. First, an organization cannot be reduced to a structure. It includes the processes that take place therein. Second, an organization cannot be limited to its prescriptions. It is the result of a continuous design process involving prescribing organizers, who aim to confine work within sets of rules, and operators, who aim to cope with the diversity of situations at work.

Thus, Fabrice Bourgeois and François Hubault argue that the activity of operators is an object that is both organized *by* the organization and a reorganizer *of* the organization. This activity aims to adapt the organization of work whenever it is found wanting and insufficient to cope with real work situations. From this point of view, activity is a resource for organizations and contributes to the work of organizations. Adelaide Nascimento, Lucie Cuvelier, Vanina Mollo, Alexandre Dicioccio and Pierre Falzon argue a similar point in the case of the construction of safety. In this field, two models confront each other: that of rule-based safety, where safety is expected to be achieved by respecting prescriptions, and that of adaptive safety, which views prescriptions as insufficient to deal with the real world, either because it escapes them or because these prescriptions are counterproductive or inefficient. These authors argue in favour of a constructed safety, combining rule-based safety and adaptive safety.

This leads to the idea of a joint evolution of human activities and organizations. According to Johann Petit and Fabien Coutarel, organizations undergo a continuous process of continuous transformation. Ergonomic interventions should assist this process, highlighting flaws in prescribed work, contributing to remedying these flaws, and setting up arenas for a debate. The goal is to achieve an adaptive organization, which operators might then be able to suit to their practices. Following the same perspective, Justine Arnoud and Pierre Falzon apply the instrumental paradigm, introduced by P. Rabardel and described in this book by Gaëtan Bourmaud, to organizations. An organization is an artifact, a human creation. In order for it to become an instrument, its 'users' must be able to both take it over and suit it to their own needs.

The constructive component of interventions

Interventions themselves are an opportunity for development, and it is this development that brings about change. This constructive aspect, advocated in several chapters, is posed as an explicit goal for ergonomic interventions.

According to Pascal Béguin, the success of a project aiming to design an artifact – be it a technical or an organizational artifact – requires a mutual learning process involving designers, on the one hand, and future users, on the other. These stakeholders will discover the needs and possibilities of one another. In the end, designers and users design more than an artifact: they design an instrument. This developmental view of design is at the heart of the approach proposed by Flore Barcellini, Laurent Van Belleghem and François Daniellou to support design projects. The issue here is to join together ergonomic work analysis, a participatory approach and the simulation of work. This approach encourages operators to take over and implement the results of a project, as well as other project stakeholders, to gain some control over these results. Indeed, the effects of this approach impact operators and designers as well as decision-makers and staff representatives.

In the context of an intervention aiming to prevent musculoskeletal disorders, Fabien Coutarel and Johann Petit point out that the development of professional activities in and through ergonomic interventions constitutes a major lever for preventive action. This development results from a combination of external margins for manoeuvre (i.e. the plasticity of the work system) and internal margins for manoeuvre (i.e. individual adaptability). The authors note that the manner in which an ergonomic intervention is conducted can lead to sustained effects that reach beyond the duration and perimeter of the intervention itself. Demonstrating a possibility to act on working conditions and on the work environment contributes to these later developments.

Several chapters offer methodologies intended to support development. Vanina Mollo and Adelaide Nascimento propose some collective reflective tools, all of which are based on confronting operators to the reality of their activity and supporting exchanges that focus on these confrontations. These tools produce two types of results. First, they improve the effectiveness of productive activity, and allow operators to reach better solutions as well as a greater diversity of solutions. Second, they increase the ability of collectives to debate about and cope with situations that they have not yet come across. However, some prerequisites need to be met for these tools to be effective: there is a need to take into account real work activity, a need for a perennial collective, a genuine possibility of action and the involvement of management.

Similarly, the methodology of co-constructive analysis proposed by Justine Arnoud and Pierre Falzon aims to support confrontations between the practices of operators working in different professions – and, in the example described in Chapter 16, working on remote sites – contributing to the same process. Crossed visits were organized at significant times during the activity. Observing the work of others and using this newfound visibility to support exchanges between professionals encouraged an awareness of mutual dependencies, as well as the construction of a transverse collective.

Bénédicte Six-Touchard and Pierre Falzon propose a training course in work analysis as a means to assist the work of experienced operators who are required to tutor novices, as well as the work of the novices themselves. The goal is to help the experienced operators to convert their incorporated knowledge into knowledge that can be verbalized and passed on. It is also to help the novices to convert their general ability to learn into an ability to conceptualize based on experience. The training course in self-analysis of work should allow them to acquire productive, functional skills, and constructive, metafunctional skills.

References

Argyris, C., and Schön, D. (1978). *Organizational learning: a theory of action perspective*. New York: Addison-Wesley.

Becker, G. S. (1964). *Human capital: a theoretical and empirical analysis, with special reference to education*. Chicago, IL: University of Chicago Press.

de Montmollin, M. (1993). Compétences, charge mentale, stress: peut-on parler de santé "cognitive"? Presented at 28th Congress of SELF, Geneva, Switzerland, September.

de Terssac, G., and Lompré, N. (1996). Pratiques organisationnelles dans les ensembles productifs: essai d'interprétation. In J. C. Spérandio (Ed.), *L'ergonomie face aux changements technologiques et organisationnels du travail humain* (pp. 251–66). Toulouse: Octarès.

Falzon, P. (1994). Les activités méta-fonctionnelles et leur assistance. *Le Travail Humain*, 57(1), 1–23.

Falzon, P. (1996). Des objectifs de l'ergonomie. In F. Daniellou (Ed.), *L'ergonomie en quête de ses principes*. Toulouse: Octarès.

Falzon, P. (2005). Ergonomics, knowledge development and the design of enabling environments? In Humanizing Work and Work Environment Conference (HWWE 2005), Guwahati, India, December.

Falzon, P., and Mas, L. (2007). Les objectifs de l'ergonomie et les objectifs des ergonomes. In M. Zouinar, G. Valléry, and M. C. Le Port (Eds.), *Ergonomie des produits et des services, XXXXII° Congrès de la SELF*. Toulouse: Octarès.

Falzon, P., and Mollo, V. (2009). Para uma ergonomia construtiva: as condições para um trabalho capacitante. *Laboreal*, 5(1), 61–69.

Falzon, P., and Teiger, C. (1995). Construire l'activité. *Performances Humaines and Techniques*, special issue, 34–39.

Faverge, J. M., Leplat, J., and Guiguet, B. (1958). *L'adaptation de la machine à l'homme*. Paris: PUF.

Leplat, J. (1977). Les facteurs déterminants la charge de travail. *Le Travail Humain*, 40(2), 195–202.

Leplat, J. (2000). *L'analyse psychologique de l'activité en ergonomie*. Toulouse: Octarès.

Pavageau, P., Nascimento, A., and Falzon, P. (2007). Les risques d'exclusion dans un contexte de transformation organisationnelle. *Pistes*, 9(2). http://www.pistes.uqam.ca.
Sen, A. (2009). *The idea of justice*. Cambridge, MA: Harvard University Press.
Wisner, A. (1995). *Réflexions sur l'ergonomie (1962–1995)*. Toulouse: Octarès.

Acknowledgements

In 2006, my mandate as president of the International Ergonomics Association (IEA) was nearing its end. The IEA was celebrating its 50-year anniversary; it was the right time for reflecting on the purpose of the discipline. In my inaugural address to the congress, I proposed a constructive and developmental view of ergonomics and its goals. This collective volume expands this view. Its form is that of a manifesto, addressed to ergonomists of all countries.

The relevance of a collective work owes a lot to the dedicated support of all the persons involved in the project. This support was immediate. The goal that had been set to the authors was not to reinterpret their past activities in the light of development; rather, the goal was to put into words, to clarify the developmental project that was already underlying these activities. Contributing to this book thus provided the authors with an opportunity to formalize, and perhaps realize, the constructive orientation of their work.

The wish to maintain a guiding line and some balance in a collective work forces the coordinator to formulate some requests, sometimes harsh, for revisions and greater conciseness. The authors complied with these requests, being conscious of putting their pen at the service of a greater project, for the benefit of the community. I thank them here.

Among the authors, some have provided me with particular support. Vanina Mollo and Adelaide Nascimento carried out a critical review of some chapters. Justine Arnoud and Julien Nelson homogenized the presentation of all bibliographical references.

The translation was carried out by Julien Nelson with extensive proofreading by Fabien Coutarel, François Daniellou, Vanina Mollo and Julien Nelson.

My thanks go to all of them.

Finally, I pay homage to a colleague. It was during one of my numerous and ever-fruitful exchanges with Catherine Teiger that the very term *constructive ergonomics* was born.

Pierre Falzon

About the Editor

Pierre Falzon, PhD, is full professor of ergonomics at the Conservatoire National des Arts et Métiers, an academic institution located in Paris, France. He has published or edited several books about ergonomics in French, English, Portuguese and Spanish. Dr. Falzon served as president of the International Ergonomics Association from 2003 to 2006. He has been awarded the Distinguished International Colleague Award by the Human Factors and Ergonomics Society (USA).

Contributors

Justine Arnoud, PhD, is a postdoctoral research assistant at the Institute of Labour Economics and Industrial Sociology (LEST, CNRS) of Aix-Marseille University. She holds a master's in economics and management from Universities of Paris 1 La Sorbonne and Rennes 1, and a specialist master's (MS) in human and organizations management from ESCP Europe. Her PhD in ergonomics from Conservatoire National des Arts et Métiers focused on enabling environments/organizations and on potential links between ergonomics and management.

Flore Barcellini, PhD, is an assistant professor of ergonomics at the Conservatoire National des Arts et Métiers in Paris, France. Dr. Barcellini's research activities deal with cooperative work and its assistance (computer-supported cooperative work), ergonomics in design project management and online communities.

Béatrice Barthe, PhD, is an assistant professor at the University of Toulouse and member of the CLLE-LTC Research Institute (CNRS). Her research focuses on atypical work schedules (shiftwork, nightwork, and extended work periods). She investigates activities workers develop in order to maintain vigilance and to preserve health and job outcomes, i.e. individual and collective strategies of regulation, resting or napping during nightshift or work–family balancing.

Pascal Béguin, PhD, is full professor of ergonomics at the University of Lyon II (Institute for Work Studies of Lyon – Centre Max Weber, CNRS). Founder of the open access journal @ctivités and chair of the Scientific & Technical Committee « ATWAD » International Ergonomics Association, he has been invited professor at the Centre for Activity Theory and Developmental Work Research (Helsinki, Finland), the Bushfire Cooperative Research Centre (Tasmania, Australia) and the Ergonomics Centre of the Faculty of Biological Sciences (Concepción, Chile).

Fabrice Bourgeois is a consultant ergonomist and the co-manager of Concilio, France. He has published books and papers on the prevention of musculoskeletal disorders in the workplace, and on the impacts of Lean production and organizational systems on health and efficiency. He teaches ergonomics in different universities. His areas of practice concern the improvement of working conditions, the design of technical and organizational systems, the mobilization of management and retention in employment.

Gaëtan Bourmaud, PhD, is a consultant ergonomist and an associate professor of ergonomics at the Conservatoire National des Arts et Métiers in Paris, France. His main intervention domains concern the design and accessibility of artifacts and working environments, and the adaptation of work systems to people with disabilities. His PhD in ergonomics examined the design and use of artifacts, using an instrumental genesis framework.

Sandrine Caroly, PhD, is an assistant professor of ergonomics at Polytech Grenoble, France, where she heads the Risk Prevention Department and teaches the methodology of ergonomics intervention and the design of work and of work organization. Her research, conducted in the PACTE Laboratory, deals with collective activity development, notably for MSD prevention, production systems and health, and the management of risks. Dr. Caroly investigates the relation between work and health in various sectors such as services, industrial production and occupational health.

Karine Chassaing, PhD, is an assistant professor of ergonomics at the École Nationale Supérieure de Cognitique, Institut Polytechnique de Bordeaux, an engineering school located in Bordeaux, France. She conducts field research on musculoskeletal disorders prevention and gestures at work in the industry sector.

Fabien Coutarel, PhD, is an assistant professor of ergonomics at Clermont University and belongs to the Activity, Knowledge, Transmission and Education (ACTé) Laboratory. His research projects focus on one hand on interactions between health and work (musculoskeletal disorders and psychosocial risks) and organization of work and companies, and on the other hand on the methodology of ergonomics intervention.

Lucie Cuvelier, PhD, a safety engineer and ergonomist, holds a PhD in ergonomics. As an assistant professor at Paris 8 University, her research concerns processes of competence development and knowledge construction, especially in the fields of occupational health, system reliability and industrial risks management.

François Daniellou, PhD, is full professor of ergonomics at the Institut Polytechnique de Bordeaux, an engineering university. He was awarded the Triennal Outstanding Educators Award by the International Ergonomics Association in 2009. His books about activity analysis, the introduction of ergonomics in design project management and human factors in high-risk industries have been published in French, English, Spanish and Portuguese.

Sandro De Gasparo is a consultant ergonomist, associated researcher at the Analysis of Work and of Mutations of Industry and Services (ATEMIS) Laboratory and also teaches at the Ergonomics and Human Ecology Department, Sorbonne University, Paris. His intervention domains are mental health and work, management and prevention of occupational risks, services activities and performance models assessment.

Catherine Delgoulet, PhD, is an assistant professor of ergonomics at the Université Paris Descartes in Paris, France. Her main domains of research concern health and ageing at work, including vocational training. She has developed these themes within the scientific program of the Centre of Research on Experience and Age of Labor Force (CREAPT) through partnerships with industry and services companies.

Alexandre Dicioccio, PhD, works as a safety engineer in a French regional airline. His PhD in ergonomics examined the trade-off between safety and performance in the field of aircraft maintenance. He now focuses on the safety risks that are involved not only in the area of aviation, but also in medicine. His research interests are team performance, resilience, simulation and crew resource management.

Irène Gaillard, PhD, is an assistant professor at the Université de Toulouse, France. She teaches ergonomics and health and safety at work at the Institut de Promotion Supérieure du travail – Conservatoire National des Arts et Métiers (IPST-Cnam). She conducts research at the Centre d'Etudes et de Recherche: Travail, Organisation, Pouvoir (CERTOP) on the relationships between activity, organization and the meaning of work.

Corinne Gaudart, PhD, belongs to the Laboratoire Interdisciplinaire pour la Sociologie Economique (CNRS), in Paris, France. She heads the Centre de Recherche sur l'Expérience, l'Age et les Populations au Travail, a research centre associating public institutes and companies. She has recently edited a book about ageing, experience, health and working conditions. Her research focuses on the construction of work experience with age, mixing psychological, sociological and historical approaches.

François Hubault, PhD, is an assistant professor of ergonomics at the Université Paris 1 Panthéon-Sorbonne, head of the Ergonomics and Human Ecology Department, and founder of the Analysis of Work and of Mutations of Industry and Services (ATEMIS) Laboratory. Hubault served twice as president of the Société d'Ergonomie de Langue Français and as Secretary General of the Centre for Registration of European Ergonomists. His research and publications concern management activities and the stakes of the nonmaterial economy.

Élise Ledoux, PhD, is a researcher at the Institut de Recherche Robert-Sauvé en Santé et Sécurité in Montréal, Québec. Her PhD is in ergonomics, and her research interests concern the organization of work and the design of spaces, service relationships and prevention in the service sector, and the safe, competent integration of new workers.

Yannick Lémonie, PhD, is an assistant professor at the Conservatoire National des Arts et Métiers in Paris, France. His research concerns development of gesture at work, the relationships between activity and learning at work and in other domains (sport and physical education) and ergonomics of teaching–learning. He has served as vice president of the Research Association on Intervention in Sport (ARIS) since 2008.

Vanina Mollo, PhD, is an assistant professor at the Université de Toulouse, France, and member of the Centre d'Etude et de Recherche Travail-Organisation-Pouvoir (CERTOP, CNRS). Her research concerns the collective construction of safety in two main ways: the impact of collective reflective practice on skills and knowledge development and on safety, and patient participation to healthcare safety.

Adelaide Nascimento, PhD, is an assistant professor of ergonomics at the Conservatoire National des Arts et Métiers in Paris, France. Her PhD addressed the issues of patient safety and safety culture in radiotherapy. She developed the differential acceptability assessments (DA2) method to analyze the acceptability of situation-specific deviations and assess safety cultures. She is currently conducting research on constructive safety, i.e. how to maintain safety while adapting general procedures to local situations.

Johann Petit, PhD, is an assistant professor of ergonomics at the Polytechnic Institute of Bordeaux, France. His research projects focus on links between organization and health. More precisely, he is interested in the practice of ergonomists in organizational design, and in the influence of the forms of organization and of organizational changes on work.

Sophie Prunier-Poulmaire, PhD, is an assistant professor of ergonomics at the Université Paris Ouest Nanterre-La Défense. Her research interests concern working conditions and the temporal organization of work, as well as health and safety, particularly in the service and retail sectors. She has also contributed to research in design ergonomics. She is affiliated with the International Society for Working Time and Health Research.

Bénédicte Six-Touchard, PhD, works as a consultant in ergonomics (risk prevention, handicap, architecture and project management). She also teaches ergonomics at the Conservatoire National des Arts et Métiers in Normandy, France. Her PhD addresses the relationships between ergonomics and vocational training and the use of activity analysis as a training tool, allowing workers to become aware of their own skills and to better develop them.

Cathy Toupin, PhD, is an assistant professor of ergonomics at the University Paris 8 (France). Her research aims to appreciate to what extent, and in which conditions, experience allows one to overcome the difficulties of work in atypical schedules (health disorders, fatigue accumulation and conditions of realization of tasks at night) in order to protect the health of the operators and develop their efficiency.

Laurent Van Belleghem is director of Realwork SAS (Paris) and associate professor of ergonomics at the Conservatoire National des Arts et Métiers (Paris). His intervention domains deal with human development at work, ergonomics in design project management and contributions to company strategies.

Christine Vidal-Gomel, PhD, is an assistant professor in the Department of Education at the University of Nantes. Her PhD research in psychology and ergonomics examined the evolution of operators' skills in managing occupational risks. Her current research investigates professional learning and development at work, still focusing on operators' skills in occupational risk management, in the purpose of devising training programs.

section one

Resources and conditions for development

chapter one

The development of skills

A condition for the construction of health and performance at work

Catherine Delgoulet and Christine Vidal-Gomel

Contents

Ergonomics has always been concerned with the development of operators, although this has not been at the heart of its research issues (Waterson
et al., 2012). In France, it has only been since the 1990s that ergonomics
has truly tackled this issue, in particular through the goals of transforming and designing work tools and work situations – typically, in the case
of situations and tools for which it was compulsory to take into account
the professional skills involved, as well as their potential transfer and the
related training practices. Hence, the concept of *skill* gradually became

essential in order to account for the fact that the activity of operators is neither haphazard nor completely predictable, and that it cannot be reduced to a list of instant behaviours. More recently, ergonomics has made this one of its main goals, through the concept of 'enabling environments' (Falzon, 2008).

Following this view, we will describe in detail the relationship between skills and the two key dimensions of ergonomics – health and performance – and the theoretical stance underlying this thesis. We will then describe how ergonomics can contribute, on the one hand, to identifying the situational conditions that are necessary for the design of work environments, and on the other hand, to highlighting the organizational and technical choices that hinder the development of skills through work and for work. Finally, we will describe at length the contribution of ergonomics to the design of training systems, viewed as a specific kind of systems aiming to develop skills, but also, indirectly, to transform work. In concluding, we will return to the scientific and social stakes that remain in order to support and develop the approach of constructive ergonomics.

Skills as a vector of health and performance

In ergonomics, health is defined as 'an ongoing tentative construction … that can be hindered by situations encountered by the subject' (Daniellou, 2006). Health is dynamically constructed, through interactions with the material, economic and social environment. Work is an important aspect of this construction. It may be a source of adverse effects on health, but it also provides possibilities for physical protection, self-fulfilment, social recognition and the development of skills (Doppler, 2004). All of these dimensions contribute to setting this dynamic balance, and to achieving the professional goals of the operator.

The preservation of health through the development of skills

Montmollin (1993) is probably the first author to have highlighted the relationship between the cognitive aspects of activity and health. Montmollin opposed health with 'cognitive misery', stating that the dynamic balance that is implied in health also requires the development of skills. Following Vygotski (1934/1986), it can be reasoned that the set of skills acquired by a subject constitute so many psychological instruments, fostering the development of higher mental functions and of capacities to apprehend the outside world. Thus, at work, health occupies an intermediary space – between 'low load', which is a potential source of boredom, disinvestment and fatigue in the face of idleness, and 'overload', leading to overwork and health hazards. The challenges here involve creating possibilities for development (Montmollin, 1993). A recent inquiry on the relations between age, health and work (Molinié, 2005) identified a statistical relationship between the fact of 'having or not having a job which makes it possible to learn', the feeling of 'being capable of remaining in employment until retirement', and the presence of health disorders and signs of wear and tear. These results emphasize the role of learning, and more broadly, development of the subject, in the construction of health at work. They show the importance of being able to learn on an everyday basis, within one's professional activity.

The relations between skills and health can be viewed from many different angles. Identifying, understanding, anticipating and dealing with hazardous situations – all of these rely on using individual and collective skills (Vidal-Gomel and Samurçay, 2002; Marc and Rogalski, 2008), whether the hazards involved are hazards for oneself, for one's teammates, for a technical system, or more broadly, for the users of a service or for the general population. Operators' skills are also involved in preventing occupational diseases, and more generally in protecting their health. In the meat processing industry, mastering how to sharpen a knife implies constructing a representation of the edge of the blade, and contributes to the prevention of musculoskeletal disorders (MSDs) (Chatigny and Vézina, 1995). In the steel industry, elder operators in charge of process control are able to diagnose the quality of steel based on the collection of sensory information, thanks to the skills they have developed in past experiences on the production line. These allow them to anticipate drifts in production with a greater time span than younger operators. This suppresses the need for them to act in an emergency, which is, to them, a source of added fatigue (Pueyo et al., 2011).

Following this, skills and their development are interesting to ergonomists for two reasons – performance and health.

Acting and understanding: conditions for constructing performance?

Skills are partly developed in work situations, through the activity deployed in the real world to complete a task. One can distinguish two dimensions of activity (Samurçay and Rabardel, 2004): *productive activity*, which is geared toward the production of goods and services, and *constructive activity*, which takes part in the development of the subject. Constructive activity focuses on the experience of the subject, as a result of productive activity. These two dimensions cannot be entirely set apart and yet are different from one another. Constructive activities are related both to carrying out an action in the here and now, to metafunctional activities (Falzon, 1994), whose goal is notably to construct tools to be used in the future, and to reflective activities (Schön, 1983). These various facets of constructive activities foster awareness (Piaget, 1974/1976), and thus conceptualization – understanding the determinants of the success or the failure of an operation.

Conceptualization is elaborated over time, within professional environments that are more or less facilitative. It provides operators with new means to complete tasks and to construct health at work, e.g. through an extension of 'fields covered by their representations' and of the 'temporal field', allowing anticipation; through a greater 'resistance to disruptions' in work situations – disruptions that will be neutralized, partially compensated or integrated within the realm of 'normal work' (Vidal-Gomel and Rogalski, 2007; Weill-Fassina, 2012). Analyzing these processes and the conditions of their maturation highlights a stake of ergonomic intervention: to contribute to the design of enabling environments (Falzon, 2008) that preserve the health of operators, allow them to integrate the company and to remain within, while fostering learning and providing the means to carry out tasks.

Thus, studying the processes of the development of skills also implies studying the issue of performance from the point of view of the organization (production, quality, safety, time, etc.), but also from that of people eager to provide 'quality work', work that is carefully performed (Clot, 2012). In this case, performance is also related to health, as revealed in current work on psychosocial hazards (Gollac and Bodier, 2011). As ergonomists, we do not view performance solely in a positive light, as a level of mastery, expertise or success in carrying out a task as it is prescribed or expected to be performed. The errors made or difficulties encountered when striving to achieve the requirements of high-quality work are just as interesting as the successes of an expert. The skills underlying successes or failures are just as worthy of our attention in order to understand to what extent, and how, it is possible (or impossible) to achieve a set of goals, while preserving or strengthening health (or not doing so) over time (Vidal-Gomel and Samurçay, 2002; Weill-Fassina and Pastré, 2004). It is by confronting these skills with real-world work situations that analysis will allow us to identify the ingredients necessary to building the conditions of sustained performance, in spite of the uncertainties that are part of any work situation, and in some cases, of the impossibilities that are created in these situations.

In our view, work so far has remained in the background of discussions regarding the relationship between health, skill and performance that we wish to present. We will address this now, highlighting the possibilities or impossibilities of development that are fostered by work environments.

Work environments for the development of skills

Temporal and situational aspects of work are crucial in the construction of skills, whose development is neither random nor predetermined. This development is strongly related to the real-world work situations in which operators evolve (Weill-Fassina, 2012). These situations may or may not be 'situations of potential development', i.e. situations that fulfill a set of conditions necessary to 'starting off and supporting the processes of skills development within individuals or groups of individuals' (Mayen, 1999, p. 66, our translation).

Favourable conditions: Potential situations of development and mediation

The first of these conditions relates to the connection between the actual skills of a person or collective of persons, and the features of the professional situation they are involved in. This connection can be described from the idea of an 'envelope of situations' (Rogalski et al., 2002), which corresponds to the zone of proximal development of individuals and of the collective (Vygotski, 1934/1986). This envelope comprises situations that, in order to be apprehended, require some learning and the implementation of mediations that can be based on tools (e.g. support tools), on peers or on other people (e.g. tutors, trainers). Situations that are included in this envelope are well-known, even routine situations. Situations that are outside of it cannot be considered, taking into account the skills that have been acquired so far.

Other conditions are related to the features of work situations themselves, which may be a source of mediations (Samurçay and Rogalski, 1998). The work organization that is implemented, the type of tasks or missions that need to be completed, the actual conditions in which they are carried out, and the ways in which people are employed and their professional histories are managed within organizations will colour and drive the development of skills, or hinder it.

Highlighting the fact that skills are developed through the situated realization of activity shows the importance of the processes of mediation based on work situations, but also on peers. These two types of processes are strongly interdependent. Let us once again take the example of process controllers in the steel industry (Pueyo et al., 2011). It is because both the work organization and the conditions of activity realization allow elder workers to carry out their tasks with sufficient decision latitude in temporal and operational terms (e.g. leaving the control room to go examine steel production with the naked eye) that these operators have been able to preserve, transfer and utilize the skills related to evaluating the quality of steel, which they had acquired previously by occupying positions on the production line. Peer-based mediation (Delgoulet et al., 2012b) can take shape in action itself (with many people carrying out an action together), in the 'off-peak' periods of work (e.g. in commuting periods or times of equipment maintenance), following incidents, or during periods occurring 'at the edge of work' (e.g. break times). Mediation is not always one-directional, but can also allow mutual improvement. The skills acquired are related to technical knowledge and know-how, but also to more labile and implicit knowledge, that is specific to a community of practice (e.g. ethical considerations, work values, etc.) or to more 'strategic' knowledge, such as 'knowing how to manage operational leeway' (Teiger, 1993), 'know-how of caution' (Cru and Dejours, 1983) or handing down 'care' (Gaudart and Thébault, 2012). These make it possible to protect oneself from hazards to one's own – or to other people's – health, without doing away with task goals.

Weill-Fassina (2012) noted that these mediations, both situational and peer based, operate whenever the (situation-specific) room for manoeuvre is sufficient to allow operators to develop responses that are suited to real-world situations. Constructing this room for manoeuvre depends in part on situated activity, at both the individual and collective levels. More broadly, it depends on the activities of designing work situations, work tools and work environments. Ergonomists can then contribute to this construction, by taking into account the three dimensions that define 'enabling environments' (Falzon, 2008).

Detrimental conditions: Between strong prescriptions and uncertainty

Yet, some work situations do not foster this interplay between productive and constructive activity. The connection between the two can thus be endangered in two opposite types of situations (Weill-Fassina, 2012): on the one hand, when rules and prescriptions are too numerous, too strict or contradictory, and when they constrain human activity and its regulations, and on the other hand, when no rules exist, generating excessive uncertainty with respect to the available skills and to potentially grave consequences for oneself and for the sociotechnical systems. Furthermore, work situations that are overly routinized, which have led to the elaboration of automatic behaviour and do not allow the introduction of controlled variation in the repetition of this behaviour, are well known to be situations that are no longer conducive to development. On the contrary, they stifle and hinder it (Leplat, 2005). Other features of work situations, such as the temporal organization of tasks, can be blamed. Drawing once again from the meat processing industry, sharpening a knife is an essential skill to prevent musculoskeletal disorders (MSDs), but it cannot be acquired in real-world work situations, because no time is allocated to completing this task (Chatigny and Vézina, 1995).

Furthermore, the constructive dimensions of human activity – whose importance for professional development we have stressed above – can be minimized when little space is devoted to them in work. Whereas some types of organizations – such as 'learning organizations' and 'Lean production systems' – imply a continuous development of operators, the rise of market constraints in many different fields of professional activity (e.g. through subcontracting, outsourcing or working in networks of subsidiaries), the diffusion of information and communication technologies, automation, as well as the increase in time constraints, all lead to a contraction of time periods that are not viewed as immediately related to production, at the cost of spaces for learning and reflection. Time that might be devoted to welcoming new recruits or to collective exchanges on work practices is reduced, or even prohibited (Delgoulet et al., 2012b), leaving each individual faced with his or her own responsibilities, doubts and difficulties at work, leading to potentially detrimental effects on production, and on the health or safety of people and systems. More broadly, it is often a 'sequence of determinants' that can be identified as a source of hindrance (Cloutier et al., 2012): from the macro-organizational level (development of low-security jobs, failure to renew employees who retire) to the level of work collectives (work overload, individualized performance assessment), including the local level of a single department or production workshop (high turnover, split shifts with no overlap). All of these factors contribute to weakening the peer-based mediation process.

Therefore, all work situations cannot be seen as 'situations of potential development'. The ergonomist's action relates both to identifying these situations and to transforming them, in order to contribute to the design of enabling environments. However, this is not enough: it is not always possible, or even desirable, to learn in the course of one's work. Therefore, initial and continuing training remains an essential lever of development, which ergonomic action should contribute to.

Designing training systems: The development and transformation of work

As early as the 1950s, ergonomic work analysis took an interest in issues of adult training (Lacomblez, 2001). Interactions between ergonomics and training design have continued since, although some sources of misunderstanding remain. These can be explained by examining the initial goals of ergonomics, which sought first and foremost to suit the job to the worker, whereas training design might be construed as having the opposite goal. However, since the 1990s, some ergonomists have been developing a different approach to ergonomics in occupational training (Lacomblez et al., 2007).

Following this approach, the contribution of ergonomics to training design fosters the development of new skills. The analysis of work (in its productive and constructive dimensions) is viewed as a primary drive in the design of training programs, or can constitute a tool and an object for training operators and transforming work situations (Delgoulet et al., 2012a). It can also be argued that what is at stake in training, from the point of view of both the activity of trainees and that of trainers, relates to ergonomic issues of health and performance.

Work analysis as a means to construct the external consistency of a training system

Thus, it has become commonplace to claim that analyzing work activity is 'a prerequisite for training' (de Montmollin, 1974, our translation), in particular because the real-world work, the needs and the difficulties of operators often remain little known. As a consequence, training systems can be designed based on idealized work situations (organizational, material and physical work conditions, level of involvement of customers or users, etc.), all of which can hinder, or even block, learning processes or can make them ineffectual in professional situations (Delgoulet, 2001; Santos and Lacomblez, 2007).

Here, we would like to emphasize another contribution of ergonomics to training design, which concerns work situations that do not make it possible to learn a trade well enough. Work analysis in real-world situations makes it possible to uncover these impossibilities, and to design situations of training by transforming the features of these real-world situations so that they might support the acquisition of new skills.

Such is the case for work situations that are associated with a high accident risk, or whose occurrence is too rare for in situ training to be a credible possibility. In these cases, simulation-based training is favoured, e.g. in the nuclear and aviation industries, or in the medical field. Such high-risk systems have a long tradition of investing in operator training, but it may not be the case for other fields, notably when operator tasks are viewed as being simple a priori. The activity of delivering ready-for-use concrete is one such case. Thus, hazard prevention and incident recovery can lead only to in-class training sessions lasting only a few hours; in contrast, the procedures involved cannot be learnt in real-world situations because of the related risks, nor can they be handed down by peers, because they are little known. Thus, experienced operators who had taken part in these training programs did not know the procedures for recovering these incidents, which could prove fatal (Vidal-Gomel et al., 2009).

From there, work analysis becomes an essential tool to identify the difficulties of operators in real-world work situations, in order to ensure the external consistency of training programs (Delgoulet, 2001) and to design training systems that make it possible to achieve levels of performance that are satisfactory to both organizations and stakeholders.

Work analysis as a tool and an object for training

This conceptual turn has also made it possible to apprehend the relationship with training considering, on the one hand, work analysis as an object for training, and on the other hand, ergonomic interventions as formative actions (Dugué et al., 2010). These two aspects are also justified by the fact that ergonomists, at the turn of the twenty-first century, are subject to an increasingly pressing need to go beyond a 'classical' process of diagnosis and its related recommendations. The rise of ergonomic interventions coupled with design projects, the need to accompany transformative action through the implementation of participatory design practices, and the wish to ensure the sustainability of ergonomic action have all led ergonomists increasingly toward the field of training design.

The stake here is to depart, once and for all, from the position of the ergonomist as an expert of the situations and solutions that should be constructed, so as to accompany various audiences toward a practice of analyzing work activity, as a part of training systems geared by and for action (Teiger and Montreuil, 1996; Six-Touchard and Falzon, this volume). Thus, 'relay' people within the company can be sensitized or trained (e.g. through close supervision by line management) in order to ensure the continuity of ergonomic action, which is primed in the ergonomic intervention, usually with no guarantee of sustainability. This is the case for the participatory approaches that may be used for the sustained prevention of health disorders (Gaudart et al., 2012).

These training programs aim to transform the initial representations of trainees, by emphasizing the value of real-world work as opposed to prescribed work, by opening the space of possibilities in the interpretation of situations (from a self-centric approach to a collective and shared approach). Beyond this work on representations, these training programs all claim to contribute to transforming representations of work situations, by providing stakeholders of these situations with suitable tools. In this case, training programs and ergonomic interventions are intertwined and feed off one another.

Analyzing activity during training programs in order to strengthen their internal consistency

Finally, activity analysis during a training program can follow multiple goals: improving existing training systems, designing tools for training stakeholders, or improving the initial or continued education of trainers, as well as their work conditions.

Analyzing the learning activity of trainees can reveal mismatches between training situations and the characteristics of these trainees, specific difficulties related to learning, or alternately, the diversity of strategies involved in learning (Cau-Bareille et al., 2012). Most studies emphasize the cognitive aspects of activity. Some rare studies have focused on its conative, i.e. affective and motivational, aspects. Indeed, some studies have shown that training could be a source of anxiety for elder operators, notably when it is intended to accompany deep changes in work – e.g. computerization of company services or changes in management (Delgoulet and Marquié, 2002). Following the same view, Santos and Lacomblez (2007) have identified in trainee fishermen a fear of losing occupational knowledge that they had acquired previously, when attempting to learn a new way of using one of their everyday tools. Thus, analyzing the activity of learning in its cognitive and conative dimensions has led to the identification of major obstacles to learning, and to designing training situations that will allow these obstacles to be countered or circumvented.

The design, evaluation or transformation of learning tools is often an opportunity to highlight the need to jointly address the activities of trainees and trainers (Six-Touchard and Falzon, this volume). The trainer's tools are a source of constraints for the work of learning that is done by the trainees themselves. Similarly, a tool that is intended for trainees will be off-target if it does not take into account the work of trainers and how it is effectively carried out (Vidal-Gomel et al., 2012). Thus, during a program aiming to train operators in automotive maintenance to deal with car breakdowns, trainers completed the insufficient resources that had been provided to them by using the experience of breakdowns that had been acquired by trainee garage operators (Anastassova and Burkhardt, 2009).

Finally, from the end of the 1990s onward, many authors have used the contributions of ergonomics to analyze the work of teachers, thus breaking away from a tradition that considered teaching from a strictly pedagogical point of view. However, although the activity of schoolteachers has been the focus of much research work, the activity of trainers, instructors, tutors, etc. working in professional domains is much less well known (Olry and Vidal-Gomel, 2011). An in-depth debate aiming to improve their work conditions, while remaining mindful of their health and safety, is sorely needed. In the current context of economic pressure, evolutions in management practices and technology, professions and the conditions of professional activity are transformed. Trainers are involved in this change in two key ways: their profession leads them to accompany these mutations and changes of work, and they are, themselves, subjected to them (Tourmen and Prévost, 2010). In this case, activity analysis highlights the new tasks and missions assigned to trainers, as well as the difficulties they encounter and their impact on health (Delgoulet, 2012). Amongst the difficulties that have been identified in this way, the following can be noted: the diversity and concurrence of multiple responsibilities for some trainers, the variety of tasks which they carry out, the pressures related to time management, and the long workdays. These studies also highlight the existence of subpathological disorders, intense feelings of fatigue, as well as risks of MSDs. All of these elements question the true possibilities that these operators have for professional development.

Discussion and conclusion

In this chapter, our goal has been to show how ergonomics contributes to the development of operators by supporting the development of their skills at work and in training. This, in turn, contributes to the preservation of their health, their safety, and more broadly, to the performance of the sociotechnical systems they are a part of. Over the course of our analysis, the classical ergonomic topics of the design and transformation of work situations have crossed the path of development (Béguin and Cerf, 2004). Four reasons can be found for this:

- The development of operators, through the construction and consolidation of their skills, is closely related to their health.
- This health and this intelligence of operators, and the operative trade-offs they are able to make in the face of unforeseen events in work situations, depending on existing margins of manoeuvre, are a factor of performance.
- The activity of trainers, the activity of trainees and the analysis of these activities can nourish the work of designers by contributing to the process of designing future tools and situations for training. They can also foster mutual learning between designers and operators/future users.
- An ergonomic approach to training renews the debate on design domain (e.g. organization or workplace design), in the sense that its goal is, once again, to construct situations of potential development that will allow the activity of operators and trainers to deploy.

These are, no doubt, new issues and new research topics that are opening up for ergonomists – but also new challenges that face the theoretical and methodological frameworks that they use, as well as the social context of their action, which seems, sometimes, to exhibit some contradictions.

Thus, the early work that has focused on the activity of trainers has suggested the usefulness of work analysis as a tool to improve their work conditions, their health and their safety – and also to help develop their skills. This work also stresses the limitations of the theoretical and methodological principles used in ergonomics to account in total for a profession whose conditions of work and employment are widely recognized as being poorly accessible, very varied and short-lived (Chatigny and Vézina, 2008; Delgoulet, 2012). The isolation introduced by poor job security further accentuates the labile character of this profession – where, for example, the boundaries between life at work and outside of work are difficult to bear. All this stresses news challenges to ergonomics interventions.

Finally, although national and international institutions have set the goal of learning and development throughout life, the actual contents of training programs are not always relevant to fostering the development of skills. Furthermore, the current context of intensification of labour – which may continue in the years to come – challenges training systems and the possibilities for development in work situations. Indeed, the time that can be devoted to constructive and metafunctional activities, including during training, tends to become shorter. The long-term acquisition of experience, which supports expertise and the preservation of health, has been made random by the popularization of low-security jobs. Organizational systems and evaluation systems have led to the individualization of work and to the deterioration of solidarity in work collectives. These are situation-based and peer-based forms of mediation that have weakened or disappeared, and conditions that constitute hindrances to the development of operators. All these are issues that research and interventions in ergonomics should contribute to identifying and to understanding. This will allow ergonomists to act in a constructive perspective, holding together the goals of health and performance.

References

Anastassova, M., and Burkhardt, J. M. (2009). Automotive technicians' training as community-of-practice: implementation for the design of an augmented reality teaching aid. *Applied Ergonomics*, 40(4), 713–721.

Béguin, P., and Cerf, M. (2004). Formes et enjeux de l'analyse de l'activité pour la conception des systèmes de travail. *@ctivités*, 1(1), 54–71. Retrieved from http://www.activites.org/v1n1/vol1num1.book.pdf.

Cau-Bareille, D., Gaudart, C., and Delgoulet, C. (2012). Training, age and technological change: difficulties associated with age, the design of tools, and the organization of work? *Work*, 41(2), 127–141.

Chatigny, C., and Vézina, N. (1995). Analyse du travail et apprentissage d'une tâche complexe: étude de l'affilage du couteau dans un abattoir. *Le Travail Humain*, 58(3), 229–252.

Chatigny, C., and Vézina, N. (2008). L'analyse ergonomique de l'activité de travail: un outil pour développer les dispositifs de formation et d'enseignement. In Y. Lenoir (Ed.), *Didactique professionnelle et didactiques disciplinaires en débat* (pp. 127–159). Toulouse: Octarès.

Clot, Y. (2012). Le travail soigné, ressort pour une nouvelle entreprise. *La nouvelle revue du travail*, 1. Retrieved from http://nrt.revues.org/108?lang=en.

Cloutier, E., Ledoux, E., and Fournier, P. S. (2012). Knowledge transmission in light of recent transformations in the workplace. *Industrial Relations*, 67(2), 304–324.

Cru, D., and Dejours, C. (1983). Les savoir-faire de prudence dans les métiers du bâtiment. *Cahiers Médico-Sociaux*, 3, 239–247.

Daniellou, F. (2006). Epistemological issues about ergonomics and human factors. In W. Karwoski (Ed.), *International encyclopedia of ergonomics and human factors* (pp. 43–47). Boca Raton, FL: CRC Press.

Delgoulet, C. (2001). La construction des liens entre situations de travail et situations d'apprentissage dans la formation professionnelle. *Pistes*, 3(2). Retrieved from http://www.pistes.uqam.ca.

Delgoulet, C. (2012). Being a trainer in the French vocational training system: a case study on job status and working conditions related to perceived health. *Work*, 41(Suppl. 1), 5203–5209.

Delgoulet, C., Cau-Bareille, D., Chatigny, C., Gaudart, C., Santos, M., and Vidal-Gomel, C. (2012a). Ergonomic analysis on work activity and training. *Work*, 41(2), 111–114.

Delgoulet, C., Gaudart, C., and Chassaing, K. (2012b). Entering the workforce and on-the-job skills acquisition in the construction sector. *Work*, 41(2), 155–164.

Delgoulet, C., and Marquié, J. C. (2002). Age differences in learning maintenance skills: a field study. *Experimental Aging Research*, 28(1), 25–37.

Doppler, F. (2004). Travail et santé. In P. Falzon (Ed.), *Ergonomie* (pp. 69–82). Paris: PUF.

Dugué, B., Petit, J., and Daniellou, F. (2010). L'intervention ergonomique comme acte pédagogique. *Pistes*, 12(3). Retrieved from http://www.pistes.uqam.ca/v12n3/articles/v12n3a2.htm.

Falzon, P. (1994). Les activités méta-fonctionnelles et leur assistance. *Le Travail Humain*, 57(1), 1–23.

Falzon, P. (2008). Enabling safety: issues in design and continuous design. *Cognition, Technology and Work*, 10(1), 7–14.

Gaudart, C., Petit, J., Dugué, B., Daniellou, F., Davezies, P., and Théry, L. (2012). Impacting working conditions through trade union training. *Work*, 41(2), 165–175.

Gaudart, C., and Thébault, J. (2012). La place du care dans la transmission des savoirs professionnels entre anciens et nouveaux à l'hôpital. *Relations industrielles*, 67(2), 242–262.

Gollac, M., and Bodier, M. (2011). *Mesurer les facteurs psychosociaux de risque au travail pour les maîtriser*. Retrieved from http://www.travailler-mieux.gouv.fr/Mesurer-les-facteurs-psychosociaux.html.

Lacomblez, M. (2001). Analyse du travail et élaboration des programmes de formation professionnelle. *Relations Industrielles*, 56(3), 543–578.

Lacomblez, M., Bellemare, M., Chatigny, C., Delgoulet, C., Re, A., Trudel, L., and Vasconcelos, R. (2007). Ergonomics analysis of work activity and training: basic paradigm, evolutions and challenges. In R. Pikaar, E. Koningsveld, and P. Settels (Eds.), *Meeting diversity in ergonomics* (pp. 129–142). Amsterdam: Elsevier.

Leplat, J. (2005). Les automatismes dans l'activité: pour une réhabilitation et un bon usage. *@ctivités*, 2(2), 43–68. Retrieved from http://www.activites.org/v2n2/leplat.pdf.

Marc, J., and Rogalski, J. (2008). Collective management in dynamic situations. *Cognition, Technology, and Work*, 11(4), 313–327.

Mayen, P. (1999). Des situations potentielles de développement. *Education Permanente*, 139, 65–86.

Molinié, A. F. (2005). Feeling capable of remaining in the same job until retirement? In D. Costa, W. J. A. Goedhard, and J. Ilmarinen (Eds.), *Assessment and promotion of work ability health and well-being of ageing workers* (pp. 112–117). Amsterdam: Elsevier.

Montmollin, M. de. (1974). *L'analyse du travail préalable à la formation*. Paris: Armand Colin.

Montmollin, M. de. (1993). *Ergonomie et santé*. Presented at the 28th Congress of SELF, Ergonomie et Santé, Geneva, Switzerland.

Olry, P., and Vidal-Gomel, C. (2011). Conception de formation professionnelle continue: tensions croisées et apports de l'ergonomie, de la didactique professionnelle et des pratiques d'ingénierie. *@ctivités*, 8(2), 115–149. Retrieved from http://www.activites.org/v8n2/v8n2.pdf.

Piaget, J. (1974/1976). *The grasp of consciousness*. Cambridge, MA: Harvard University Press.

Pueyo, V., Toupin, C., and Volkoff, S. (2011). The role of experience in night work: lessons from two ergonomic studies. *Applied Ergonomics*, 42(2), 251–255.

Rogalski, J., Plat, M., and Antolin-Glenn, P. (2002). Training for collective competence in rare and unpredictable situations. In N. Boreham, R. Samurçay, and M. Fischer (Eds.), *Work process knowledge* (pp. 134–147). London: Routledge.

Samurçay, R., and Rabardel, P. (2004). Modèles pour l'analyse de l'activité et des compétences: propositions. In R. Samurçay and P. Pastré (Eds.), *Recherches en didactique professionnelle* (pp. 163–180). Toulouse: Octarès.

Samurçay, R., and Rogalski, J. (1998). Exploitation didactique des situations de simulation. *Le Travail Humain*, 61(4), 333–359.

Santos, M., and Lacomblez, M. (2007). Que fait la peur d'apprendre dans la zone prochaine de développement ?. *@ctivités*, 4(2), 16–29. Retrieved from http://www.activites.org/v4n2/v4n2.pdf.

Schön, D. A. (1983). *The reflective practitioner: how professionals think in action*. New York: Basic Books.

Teiger, C. (1993). L'approche ergonomique: du travail humain à l'activité des hommes et des femmes au travail. *Education Permanente*, 116(3), 71–96.

Teiger, C., and Montreuil, S. (1996). The foundations and contribution of ergonomics work analysis in training programmes. *Safety Science*, 23(2), 81–95.

Tourmen, C., and Prévost, H. (Eds.). (2010). *Être formateur aujourd'hui. Des formateurs de l'AFPA s'interrogent sur leur métier*. Dijon, France: Raisons et Passions.

Vidal-Gomel, C., Boccara, V., Rogalski, J., and Delhomme, P. (2012). Sharing the driving-course of a same trainee between different trainers, what are the consequences? *Work*, 41(2), 205–215.

Vidal-Gomel, C., Olry, P., and Rachedi, Y. (2009). Os riscos profissionais e a sua gestão em contexto: dois objectos para um objectivo de formação comum. *Laboreal*, 2, 31–47.

Vidal-Gomel, C., and Rogalski, J. (2007). La conceptualisation et la place des concepts pragmatiques dans l'activité professionnelle et le développement des compétences. *@ctivités*, 4(1), 49–84. Retrieved from http://www.activites.org/v4n1/v4n1.pdf.

Vidal-Gomel, C., and Samurçay, R. (2002). Qualitative analysis of accidents and incidents to identify competencies: the electrical system maintenance case. *Safety Science*, 40(6), 479–500.

Vygotski, L. S. (1934/1986). *Thought and language*. Cambridge, MA: MIT Press.

Waterson, P., Falzon, P., and Barcellini, F. (2012). The recent history of the IEA: an analysis of IEA Congress presentations since 1961. *Work*, 41(Suppl. 1), 5033–5036.

Weill-Fassina, A. (2012). Le développement des compétences professionnelles au fil du temps à l'épreuve des situations de travail. In C. Gaudart, A. F. Molinié, and V. Pueyo (Eds.), *La vie professionnelle: âge, expérience et santé à l'épreuve des conditions de travail* (pp. 117–144). Toulouse: Octarès.

Weill-Fassina, A., and Pastré, P. (2004). Les compétences professionnelles et leur développement. In P. Falzon (Ed.), *Ergonomie* (pp. 213–231). Paris: PUF.

chapter two

The development of collective activity

Sandrine Caroly and Flore Barcellini

Contents

Ergonomics is interested in the collective aspects of human activity. It views these aspects being subjected to constant adjustments. These adjustments take place in the interactions between the operator and the context of his work. The goal of this chapter is to present a constructive view of the development of collective activity. This activity will be viewed as connecting the collective work that operators are involved in with the work collective that they belong to.

In the first part of this chapter, we will present the concept of work collective – as a resource for the development of health and collective work, as a resource for the development of performance, and finally, as one component of effective and efficient collective activities. In the second part, we will describe the organizational and material conditions that are crucial to developing collective activity. Finally, we will highlight the need for ergonomics to focus on the work that consists in organizing environments so that they may enable the development of this collective activity – and, as a consequence, the need to focus on the activity of the people who organize these environments.

Articulating collective work and the work collective within work activity

Collective work as a resource for performance

Collective work refers to the ways in which operators may cooperate, in a more or less effective and efficient manner, in a work situation (de Cássia Pereira Fernandes et al., 2010; De La Garza and Weill-Fassina, 1995). It is therefore defined in relation to the task that the partners of collective work are involved in, and relates to their performance with respect to achieving the goals of this task. Collective work implies processes of task allocation and knowledge sharing. These processes are related to the implementation of adjustments within the activity.

Many kinds of sociocognitive resources can foster the production of effective collective work (Caroly and Weill-Fassina, 2007; Carroll et al., 2006; Darses et al., 2001; Salembier and Zouinar, 2004; Schmidt, 2002): opportunities for operative synchronization – i.e. coordination – between participants, the construction of a common frame of reference (COFOR), reciprocal knowledge of the work of all the persons involved, and a shared reference concerning the state of progression of the process, which implies the development of situation awareness.

Operative synchronization defines the possibilities for coordination between participants involved in collective work (e.g. Darses et al., 2001). It aims to ensure the attribution of tasks between the partners of a collective work, and its organization in time (e.g. starting and stopping points, simultaneity, sequencing and rhythm of actions that need to be carried out). This coordination is never completely prespecified (e.g. through prescribed procedures). It is constructed by the partners and involves communication – both verbal and nonverbal – between them (Heath et al., 2002; Salembier and Zouinar, 2004). In particular, this communication allows the implementation of adjustment processes that ensure the effectiveness of collective work (Guerin et al., 2006). Coordination processes emerge as crucial elements for monitoring unexpected events and avoiding accident situations (de Keyser, 1991).

A second type of resource that is typically mentioned in ergonomics relates to the possibility for participants in collective work to *cognitively synchronize* with one another (e.g. Darses et al., 2001) – that is, to construct, maintain and update a set of 'shared knowledge' that allows the partners of collective work to manage the dependencies connecting their individual activities. This knowledge is based on a set of situations experienced together, and on trade-specific knowledge and beliefs that are historically and culturally constructed (Salembier and Zouinar, 2004).

Two types of knowledge appear to be essential for effective collective work:

- On the one hand, participants must be able to construct shared knowledge regarding their field of activity (technical rules, objects of the field and their properties, problem-solving procedures, etc.). This shared knowledge is also termed common frame of reference (COFOR). This framework comprises the 'functional representations shared by operators, that guide and control the activity which they carry out as a collective' (Leplat, 1991; Hoc and Carlier, 2002). In order to construct the COFOR, the agents of collective work must be able to take part in activities geared toward clarification (Baker, 2009) and explanation, in order to negotiate and construct a shared understanding of the situation (Salembier and Zouinar, 2004), but also to conceptually adjust to one another (Karsenty and Pavard, 1998) and to construct the more stable knowledge that is necessary for collective work. For example, the construction of a common frame of reference (Leplat, 1991) becomes crucial when dealing with failures of the work system and when controlling hazards.
- On the other hand, in the 'here and now' of a particular task, participants must be able to construct a representation of the current state of the situation that they are involved in (knowledge of facts related to the state of the situation, of the contributions of partners involved in the task, etc.), also known as awareness (Carroll et al., 2006; Schmidt, 2002). The construction of this awareness is supported by practices whereby participants, by cooperating and coping with their own emergencies and unforeseen events, are able to 'sense' what their colleagues are doing and to adjust their own activity accordingly (Schmidt, 2002). They must therefore 'remain sensitive to each other's conduct' (Heath et al., 2002, p. 317). However, this does not imply merely supervising the activity of one's partners, but also making visible the elements of one's own activity that might be relevant to others (Heath et al., 2002; Salembier and Zouinar, 2004; Schmidt, 2002). Therefore, the construction of awareness is not just an opportunistic process resulting from the affordances

of a situation. It also relies on the ability of the partners of collective work to recognize, interpret and understand each other's conducts and the resources that are available to them (Heath et al., 2002; Salembier and Zouinar, 2004).

The work collective: A resource for the development of health and skills

Collective work must be distinguished from the concept of the work collective. Indeed, 'all collective work does not (necessarily) involve the existence of a work collective' (Caroly, 2009). Yet, many studies tend not to make a clear distinction between what is related to the work collective versus what is related to collective work.

For ergonomics, a work collective is constructed between operators who share goals related to the realization of 'quality work' – i.e. work that is in accordance with the criteria that *they themselves* ascribe to 'effective work', and with the meaning that they give to this work. The work collective, thus created, exerts a function of protecting the individual's subjective relationship with action. This protective function expresses itself, notably, through the collective's ability to construct – or reconstruct – standards and rules to frame action, in compliance with the criteria of quality of work, to manage conflicting relationships at work, and finally, to give meaning to the work. The collective allows each of its members to access this meaning and the criteria of a 'job well done' through the 'rules of the trade' (Cru, 1988). These rules are grounded in a history that structures exchanges between people at work and fosters the mobilization of the subject in his or her own activity. In this sense, the concept of work collective refers to something more than just a collection of individuals. The collective is a part of the activity, and not just a determining factor of the work situation.

Following a constructive view, two aspects of a work collective should be taken into account. First, the work collective emerges as a resource for the development of health in a broad sense. It allows an individual to 'take care' of his or her own work and, from this point of view, contributes to individual health. Furthermore, it fosters learning and the development of skills. This development is, in itself, a driving element in the preservation of health (Delgoulet and Vidal-Gomel, this volume).

The work collective allows the preservation of the health of its members, in the sense that it ensures that the debate about work does not focus directly on issues related to personality, but on issues related to activity and work organization. The existence of a work collective leads operators to debate the meaning of their actions, and to share means of dealing with situations that are a source of conflicts of goals within their activities.

Thus, the work collective provides a number of 'gestures of the trade' and a set of ways of doing 'quality work' – in the sense that operators themselves ascribe to this term. These may help operators find, in their activity, ways and means of doing things that are suited to the situation, such that they might foster the preservation of health and the construction of a meaning of work.

More specifically, the work collective plays a part in preserving psychosocial resources (Caroly et al., 2012) and preventing musculoskeletal disorders (MSDs) (de Cássia Pereira Fernandes et al., 2010; Lemonie and Chassaing, this volume). For example, conflicts of goals are present in the everyday work of police officers (Caroly, 2011). When confronted with situations involving persons with low job security, should one arrest them at any cost to achieve the goals of the ministry, or not arrest them in order to prevent the situation from worsening further and to ensure quality of service? When the work collective defines a set of situations in which officers should not intervene (e.g. referring a homeless person to a shelter, requesting that a person go to the nearest police station to identify himself or herself, phoning a mother so that she agrees to hand her children over to her ex-husband, etc.), police officers construct a debate about the quality of work, using criteria that allow them to cope with the conflicts of values and ethical dilemmas that they encounter in real-world work situations. Conversely, in other work situations – e.g. in call centers, or Neo-Taylorian work situations in general – there are few opportunities to rely on the work collective. This leads to a heightened pressure on operators, who can no longer take care of their own work, and whose subjectivity is constantly put at odds because of the double binds that cannot be resolved by the work collective.

The work collective provides a support for innovating regarding the various ways of 'doing work'. It allows innovative forms of learning because it supports inquiry, confrontation and debate between the members of the work collective.

The need to develop collective activity

The work collective is mainly thought of as a resource for health, whereas collective work relates to the effectiveness of collective action. Yet, collective work and the work collective are linked together in the realization of the activity of subjects. Indeed, the collective is a resource for activity in the sense that it makes collective work more 'operant', for example, through the construction of shared rules to deal with outside constraints, which makes it possible to enrich the common frame of reference. Furthermore, since it supports the collective management of situations, the collective supports cooperation rather than the individual management of these situations. For example, it allows the implementation of age- or experience-related adjustments, which allows dividing up efforts, but also supports members of the group who are confronted with difficulties to achieve the goals of their tasks. Finally, collective work empowers operators at work (Clot, 2008) – and therefore makes collective work more effective.

Conversely, experiencing situations of collective work in action is a means of developing the work collective. The collective does not really exist prior to action. It is created through opportunities to act together. It relies on work situations that provide practical experiences of collective work, which are an opportunity for subjects to commit themselves to the work collective.

In order to account for this connection between the work collective and collective work, we have proposed the concept of *collective activity* (Caroly, 2009). The implementation of a collective activity pursues the goals of health, effectiveness and the development of values that are specific to that activity (i.e. the meaning of work for operators who are involved in exchanges with their colleagues about the quality of work in the trade). This collective activity allows the development of individual skills, allows these skills to complement each other in work, and enriches the liveliness of the work collective (Caroly, 2009). Thus, collective activity cannot be built solely on the basis of a sum of different individual activities, but through constant toing and froing between the activity of a subject, the implementation of collective work, and the operation of the work collective.

Collective work and the work collective are the two linchpins in the production of a high-quality collective activity. The work collective supports the development of skills, learning and the preservation of health; effective collective work supports the achievement of task goals. But this is only possible under specific conditions. Following a constructive view, ergonomics must therefore foster the development of a collective activity by acting on the organizational and material conditions of work that foster the construction of collective work and of the work collective.

Supporting the conditions to develop collective activity

Following a developmental perspective, the goal of ergonomics is to support collective activities by creating the tools and the means that are required for its development: to support the development of representations of competence and quality in the work of others, to construct spaces in which to share the criteria of the quality of work, to develop organizations supporting the processes of 'reconstruction of rules', and to design tools to support the development of resources for collective activities, via intermediary objects and technological devices to support collective activities.

Supporting the recognition of skills and quality of work in other people

Collective activity rests on knowledge of other people and on the recognition of their skills. Recognition of skill takes place not only in the vertical relationships between the hierarchy and operators. It also occurs in the horizontal relationships between operators themselves. Recognizing competence in another person, such as a colleague, is a necessity for collective work, and it enriches the work collective. Thus, this recognition can foster collective activity because it gives rise to cooperation in action and implies effectiveness in the work collective. If collective work is acknowledged and supported, this can contribute to the implementation of metafunctional activities regarding specific work situations. This can help operators to become aware of their own experience and to formalize their own skills – possibly in order to pass those skills on.

Theories of recognition (Gernet and Dejours, 2009; Dejours and Deranty, 2010; Honneth, 2012) share the idea that recognizing quality in the work of another person implies recognizing quality in the work and in the individual. The recognition of competence, as a condition for the development of collective activity, poses questions regarding how each person contributes to production and cooperation, but also on how values and standards are debated in the collective. Furthermore, assessing the competence of other people and the quality of their work are two driving forces behind constructing a relationship of mutual trust that is essential to developing a collective activity, for example, in order to communicate efficiently (Karsenty, 2011). Indeed, one cannot simply order by decree that a relationship of trust should exist. This relationship is constructed through interactions between professionals, in particular through the assessment of matches or mismatches between the expectations of the protagonists involved in work and the results obtained by their colleagues (Karsenty, 2011). Therefore, there exists a strong relationship between the possibilities of assessing competence and quality in the work of others and possibilities of constructing a relationship of mutual trust.

Ergonomics must assist the recognition of competence in the protagonists of collective activities. By using elicitation or confrontation techniques (Mollo and Nascimento, this volume), it may make it possible to formalize the skills developed by all the people involved in specific work situations – often while being at odds with the expectations of work. Thus, it leads operators to question themselves concerning what they know about each other, about their skills and their weaknesses.

Recognizing competence and quality in the work of others is a prerequisite for setting up debates regarding the criteria used to judge the quality of work, which are an essential part of collective activity.

Constructing arenas for debate in order to share the criteria for assessing the quality of work

In order to support the development of collective activity, the work collective must provide the means to support debates regarding the values, the relevant aspects of activity, and the conditions necessary to performing quality work (related to effectiveness, the preservation of health and the construction of the meaning of work). To do this, ergonomics should fuel discussions between operators focusing on the real-world work activity and on the conflicts and injunctions that are posed in specific situations. It supports a debate about the quality that is sought by each person in one's activity. It acknowledges that the criteria of work quality are related to the resources that are available and the obstacles that are encountered by everyone in their activity. Therefore, these criteria may differ depending on each person's way of thinking and acting. They do not depend on the performance criteria defined by the organization for each task, but on the real-world activity and what it requires from workers.

In a constructive approach, ergonomics must take responsibility for supporting debates about the work activity, so that members of the work collective might enter into a dialogue, regarding both the difficulties encountered at work and the internal and external resources of activity. This implies providing the collective with specific methodological tools, particularly with discussion spaces for operators from a same craft, that allow the criteria of work effectiveness and the values mobilized in a work activity to be discussed. To achieve this goal, several methods should be used more directly in order to collectively construct the criteria to judge the quality of work: methods of 'crossed auto-confrontation' and allo-confrontation or differential judgement of risk acceptability (Mollo and Falzon; Mollo and Nascimento, this volume).

The ergonomic approach can help the work collective to construct a point of view regarding what views should be defended in terms of the quality of work. Debating the criteria of the quality of work is a pre-requisite for any transformation of work situations. It makes it possible to define evolutions in the organization that should be derived from negotiations between all of the stakeholders involved (e.g. management, designers and staff representatives).

Developing an organization that supports the reelaboration of rules

The possibility of constructing or reconstructing rules shared by a collective is a crucial condition to the development of collective activity. Debating the criteria of quality of work, but also the rules of the trade and their reconstruction, plays a part in making the collective work, and in enriching it. It is instrumental to the effectiveness of collective work and exerts a protective function on the health of individuals (Cru, 1988). Reconstructions of rules by the collective aim not just to alleviate the constraints of work that derive from prescriptions of the hierarchy, but also to manage conflicts within the goals of the activity by finding ways of circumventing them to complete 'a job well done' (Caroly and Weill-Fassina, 2007). Organizational conditions that allow a confrontation with the gestures and practices of other members in the collective, and debates about the values and meaning expressed in work, are all essential to ensuring both the learning and the reconstruction of rules (Bourgeois and Hubault, this volume; Arnoud and Falzon, this volume).

Several organizational conditions must be met in order for the process of reconstruction of rules to take place:

- The rules implemented by the organization must be able to support the adjustments that are collectively put in place by the operators in order to compensate for the misgivings and the contradictions emanating from the organization. For example, when the failure to apply a prescribed rule occurs in order to manage risks, the collective must reconstruct these rules so that they can be adjusted to the real-world work activity. The operational leeway created by operators within their activity, or provided by the organization, for operators to adapt to the difficulties of their task must be completed with the operational leeway provided by the work collective (Caroly et al., 2012).
- The operational leeway provided by the organization of work must help the implementation of operative adjustments, and the construction of metarules that define the collective rules for using the prescribed rules. These are constructed through a confrontation with a varied set of experiences, and require time and the sharing of experiences.
- Individual adjustments, the attribution of tasks based on age and experience, must all be made possible within collective work. Thus, the collectively redefined rules must allow for the specificity of every person within his or her activity without hindering the production of collective work.

Constructing intermediary objects to support collective activity

The presence of intermediary objects is another condition for the development of collective activity, notably because these objects support debates about the criteria of quality of work. Boujut and Laureillard (2002) define these objects as 'a general category embracing all types of artifacts, whether physical or virtual, produced by the participants [of collaborative work] during their work'. These intermediary objects foster exchanges aiming to construct operative and cognitive synchronization, the common frame of reference and awareness. Their function is one of mediation, both for the worker in relation to his individual activity and for collective activity, by acting as support tools for a joint reflection on the situation. In this sense, intermediary objects can provide concrete grounds to a discussion and support the debates regarding the quality of work and the shared goals of operators.

In a constructive approach, these intermediary objects constitute instruments of the collective activity that ergonomists can contribute to develop. Ergonomics must make these instruments visible to agents who are not always conscious of their existence, in order to support a confrontation between them regarding the goals of production and quality. Furthermore, to support the development of collective activity, ergonomics must also take part in designing intermediary objects that will make it possible to debate various points of view and to foster controversies. For example, ergonomic approaches to design already integrate some intermediary objects of design – e.g. simulation devices – that allow the stakeholders of design to develop shared representations of the object being designed and to remember the controversies that take place surrounding the future activity – and hence the development of these activities (Barcellini et al., this volume).

Designing technical systems to support collective activity

Situations of collective work are increasingly equipped with computational tools aiming to support collective work. However, these technological systems are often designed with the goal of supporting the prescribed collective work – or even the prescribed coordination of tasks – not in terms of supporting the construction of a collective activity. For example, in most cases, these tools include a model of workflows, corresponding to a prescribed view of the coordination processes between the partners of collective work (Salembier and Zouinar, 2004). The introduction of such tools is often accompanied by an impoverished context of action, even though this context is a crucial element in understanding the activity of all the people involved and the possibilities to construct an awareness of the situation (Salembier and Zouinar, 2004) and a common frame of reference. Finally, few tools provide a direct support to construct representations focusing on the skills, roles and expertise of the other protagonists – i.e. what has been called a 'social conscience' (Barcellini et al., 2010). This restricts the possibilities for using these tools as a potential resource for the development of a work collective.

There is therefore a genuine risk of rigid work processes working deeply against collaborative work by impairing adjustments, communications and the construction of the common frame of reference. These may be made impossible because of the technical systems involved, but also because of the impossibility of accessing or recognizing the work of others, of sharing quality criteria, and because of goal conflicts that are reified within technical systems. In the medium term, it is the opportunities for the development of the work collective that may be restricted, leaving its agents with no resources available to preserve their own health.

How, then, can one design technical systems that both support collective work and allow the development of the work collective? Some of the limitations outlined above may be dealt with by proposing an approach to accompany the design projects of technical systems, so that they might aim to jointly develop these technologies, the organizations in which they are to be integrated, and finally, the future activity of operators (Barcellini et al., this volume).

One of the goals of this approach is to act on the design of a technical artifact, so as to allow the development of a collective activity:

- On the one hand, this involves contributing to the development of functions to support collective work (coordination, awareness, construction of the common frame of reference, etc.), for example, by proposing ways of representing the actions of operators, that make it possible to access both these resources and the information that is present in the environment (Salembier and Zouinar, 2004).
- On the other hand, this also involves contributing to the development of functions that support the development of the work collective. Such functions should relate to possibilities of constructing representations of other people's skills, and to functions aiming to help formalize quality criteria used by one participant or another.

However, it is quite unlikely that technology in itself would be enough to support the development of a collective activity. This approach aiming to accompany design projects must also contribute to designing the overall work situation in which the technical artifact is due to be deployed. It must therefore also support the redefinition of those rules in the organization that the technical system will contribute to transform.

Finally, the implementation of an ergonomic approach to design is an opportunity for the development of activities (Barcellini et al., this volume; Petit and Coutarel, this volume) – and therefore, of collective activities, since it should support – notably when simulations are used, the processes aiming to redefine the rules of work, and the discussions focusing on the quality criteria that are necessary for the development of a work collective.

Conclusion: The development of enabling environments and the importance of the activity of proximity management

Producing and developing a collective activity is therefore a task for operators. This task requires the development of tools and organizations to support the connection between collective work and the work collective.

More broadly, the approach we have developed in this chapter contributes to the current reflection on the design of enabling environments – in our case, that enable the development of a collective activity. As we have argued, issues surrounding rules and the redefinition of rules, and discussions focusing on the criteria of quality of work are essential in the development of a collective activity. This suggests that there is a dire need to work on the design of enabling organizations. An *enabling organization* supports the development of rules that are acceptable for an activity – both an industrial activity and a collective activity (Petit and Coutarel, this volume; Falzon, 2005; Arnoud and Falzon, this volume) – i.e. rules that make it possible to develop a quality activity accounting for goals related to the development of health (in a broad sense) and to performance, and to support a debate regarding the design of quality work (and the values related to this work).

The means that ergonomics may consider using to transform these organizations are twofold. One of these means is to help managers to be able to recognize in what ways operators reorganize their own work, and how this allows them to fuel a debate about the meaning of work and of quality work; to take this into account in the process of redesigning an organization; and to design organizations that leave room for these debates and allow for the reconstruction of these rules. A second means is to support these debates during the design process itself, through the implementation of organizational simulations that will foster controversies between operators about the rules and the meaning of work.

In other words, in order to be able to act efficiently on organizations and their design, ergonomics should now focus more on the activity of people involved in organizing work – on their constraints, their resources and their strategies – in order to make all of these elements evolve toward a better integration of collective activity in the reorganization of work. It is only then that the environment can become enabling for collective activities. This also means that the equipment used to support a collective activity (e.g. technical devices, actor networks, discussion spaces, means to pass on experience) should be investigated directly by ergonomists in the course of their interventions.

References

Baker, M. J. (2009). Argumentative interactions and the social construction of knowledge. In N. M. Mirza and A.-N. Perret-Clermont (Eds.), *Argumentation and education: theoretical foundations and practices* (pp. 127–144). Berlin: Springer Verlag.

Barcellini, F., Détienne, F., and Burkhardt, J. M. (2010). Distributed design and distributed social awareness: exploring inter-subjective dimensions of roles. In M. Lewkowicz, P. Hassanaly, M. Rodhe, and V. Wulf (Eds.), *Proceedings of the COOP'10 Conference*, Aix-en-Provènce, France. Berlin: Springer.

Boujut, J. F., and Laureillard, P. (2002). A co-operation framework for product-process integration in engineering design. *Design Studies*, 23(6), 497–513.

Caroly, S. (2009, August). Designing collective activity: a way to workers' health. Presented at Proceedings of the IEA Congress 2009, Beijing, China.

Caroly, S. (2011). Collective activity and reelaboration of rules as resources for mental health: the case of police officers. *Le Travail Humain*, 74(4), 365–389.

Caroly, S., Landry, A., Cholez, C., Davezies, P., Bellemare, M., and Poussin, N. (2012). Innovation in the occupational health physician profession requires the development of a work collective to improve the efficiency of MSD prevention. *Work: Journal of Prevention Assessment and Rehabilitation*, 41(1), 5–13.

Caroly, S., and Weill-Fassina, A. (2007). How do different approaches to collective activity in service relations call into question the plurality of ergonomic activity models? *@ctivités*, 4(1), 99–111. Retrieved from http://www.activites.org/v4n1/v4n1.pdf.

Carroll, J. M., Rosson, M. B., Convertino, G., and Ganoe, C. H. (2006). Awareness and teamwork in computer-supported collaborations. *Interacting with Computers*, 18(1), 21–46.

de Cássia Pereira Fernandes, R., Avila Assunção, A., Muniz Silvany Neto, A., and Martins Carvalho, F. (2010). Musculoskeletal disorders among workers in plastic manufacturing plants. *Revista Brasilena de Epidemiologia*, 13(1).

Clot, Y. (2008). *Travail et pouvoir d'agir*. Paris: PUF.

Cru, D. (1988). Collectif et travail de métier. In C. Dejours (Ed.), *Plaisir et souffrance dans le travail* (pp. 43–49). Paris: Editions de l'AOCIP.

Darses, F., Détienne, F., Falzon, P., and Visser, W. (2001). *COMET: a method for analysing collective design process*. Rocquencourt, France: INRIA Publications. Retrieved from http://hal.inria.fr/inria-00072330/.

de Keyser, V. (1991). Work analysis in French language ergonomics: origins and current research trends. *Ergonomics*, 34(6), 653–669.

Dejours, C., and Deranty, J. P. (2010). The centrality of work. *CRIT*, 11(2), 167–180.

De La Garza, C., and Weill-Fassina, A. (1995). Method of analysis of risk management difficulties in a collective activity: railways maintenance. *Safety Science*, 18(3), 157–180.

Falzon, P. (2005, December). Ergonomics, knowledge development and the design of enabling environments. Presented at HWWE 2005, Guwahati, India.

Gernet, I., and Dejours, C. (2009). Evaluation of work and recognition. *Nouvelle Revue de Psychosociologie*, 8(2), 27–36.

Guerin, F., Laville, A., Durrafourg, J., Daniellou, F., and Kerguelen, A. (2006). *Understanding and transforming work: the practice of ergonomics*. Lyon, France: ANACT Network Editions.

Heath, C., Svensson, M. S., Hindmarsh, J., Luff P., and Von Lehn, D. (2002). Configuring awareness. *JCSCW*, 11(1–2), 317–347.

Hoc, J. M., and Carlier, X. (2002). Role of common frame of reference in cognitive cooperation: sharing tasks between agents in air traffic control. *Cognition, Technology and Work*, 4(1), 37–47.

Honneth, A. (2012). Foreword. In S. O'Neill and N. H. Smith (Eds.), Recognition theory as social research: investigating the dynamics of social conflict (pp. 6–9). Basingstoke: Palgrave Macmillan.

Karsenty, L. (2011). Interpersonal trust and work communications. *Le Travail Humain*, 74(2), 131–155.

Karsenty, L., and Pavard, B. (1998). Various levels of context analysis in the ergonomic study of cooperative work. *Réseaux*, 6(2), 167–193.

Leplat, J. (1991). Organization of activity in collective tasks. In J. Rasmussen, B. Brehmer, and J. Leplat (Eds.), *Distributed decision making* (pp. 61–74). Chichester, UK: J. Wiley.

Mollo, V., and Falzon, P. (2004). Auto- and allo-confrontation as tools for reflective activities. *Applied Ergonomics*, 35(6), 531–540.

Salembier, P., and Zouinar, M. (2004). Mutual intelligibility and shared context: conceptual inspirations and technological reductions. *@ctivités*, 1(2), 64–85. Retrieved from http://activites.org/v1n2/salembier.pdf.

Schmidt, K. (2002). The problem with 'awareness': introductory remarks on 'awareness in CSCW'. *Journal of Computer Supported Cooperative Work*, 11(3–4), 285–298.

chapter three

The development of the psychosocial dimension of work

*Laurent Van Belleghem, Sandro De Gasparo
and Irène Gaillard*

Contents

The recent social and political emergence of 'psychosocial hazards' (PSHs) in France (Salher et al., 2007), Europe (Leka et al., 2011) and several industrialized countries (Kortum et al., 2011; Lippel and Quinlan, 2011) such as Canada (Shain, 2009), Australia (Johnstone et al., 2011) or Japan (NDCVK, 1990; Herbig and Palumbo, 1994) has questioned ergonomics to an unusual degree. Although ergonomics is, quite rightly, called upon to respond to this challenge – undoubtedly, PSHs have an impact on human work – it cannot do so without first reinstating, in its model of human activity, the psychical and social dimensions of work, which had so far been largely absent from the literature, and in so doing, reinstating in this model a theory of the acting subject.

Two hypotheses might explain this absence. The first hypothesis relates to the expectation that the psychosocial dimension of work has few functional aspects, and that it would be, at first glance, of little operational interest to ergonomics' goal of 'fitting the job to the worker' (IEA, 2006). The second hypothesis relates to the subjective nature of this psychosocial element, which would make it more relevant to the psychology of subjects than to the psychology of activity.

And yet, work situations require personal and collective mobilization. Personal mobilization operates both at the level of the *engagement* of a subject, which corresponds to an involvement of oneself in a work activity, and at the level of efficiency, which relates to the search for a response that is both operational and economical to the requirements of production work. Collective involvement relies on interactions between subjects who agree with one another regarding ways of doing things and lines of conduct to keep to.

This psychosocial mobilization is not a given fact, nor is it stable once it has been achieved. It is constantly renewed in the face of real-world situations. It gives meaning to actions that are yet to come, which in turn give meaning to actions that are immediately carried out. In other words, engagement in work is also a work of engagement. This process contributes actively to the development of the acting subject. It also contributes to the development of the social system, through the interactions that individuals undergo with one another within an organization, in order to coordinate themselves and to cooperate with one another. The dual aspect of work, both psychological and social, constitutes the driving force in the development of activity.

The emergence of psychosocial disorders at work indicates a slow-down, or even a standstill, in this dual development. As a result, the social system (notably cooperation) and the health of operators are both affected. One can then talk about a situation of 'disrupted' activity.

Hence, prevention can no longer be content with aiming to protect employees against risk factors that are solely external to the activity itself. It must also support the development of this protection, based on implementing a constructive approach to health in employees themselves and in the system. It must contribute, at the same time, to the development of the subject and to that of the social body. It is a key issue for ergonomics to understand how to support this development – not only because it contributes to the prevention of psychosocial hazards, but also because it contributes to the emancipation of individuals in work and because of work.

In this chapter, we will present in turn:

- The features of the psychosocial dimension of work that make it possible to understand the issues behind its development, but also how this development can be hindered, leading to psychosocial disorders.
- The dynamics behind the development of this psychosocial dimension, which rely on a dual process of *mobilization* of the subject when

faced with the real world, and of *sedimentation* of the results of this mobilization. This dual process allows learning through (past) actions and learning for (future) actions. It is at work both in the *acting subject* and in the *social system* composed of the various actors engaged in the work organization.

- The stake of the ergonomic intervention is to convert the development of the psychosocial function into an operational goal. The aim is to do away with a strictly preventive approach of risks, which often aims to protect operators from intangible external nuisances, and to reach for a constructive approach of human activity, which relies on optimal possibilities for development and for the mobilization of skills in individuals and work collectives. The simulation of work, but also the implementation of spaces to discuss and debate real-world work, should contribute to this.

The psychosocial dimension of work: A forgotten dimension

From psychosocial hazards to the psychosocial dimension of work

Psychosocial risks* (PSRs) are poorly named. Indeed, the psychosocial dimension of work is not a risk in itself. If we are to talk about the adverse consequences of work on employees, it would be preferable to talk about psychosocial disorders (PSDs) (Van Belleghem and De Gasparo, 2014) just as one talks, for example, about musculoskeletal disorders (MSDs). PSDs can then be defined in this way: they refer to a set of symptoms (stress, ill-being, restlessness, tension, etc.) that can develop into more severe forms (anxiety, suffering, burnout, depression, somatization) and lead to specific types of behaviour (aggression, violent behaviour, addictive behaviour, harassment) that affect the intimate sphere of the worker or his or her relations with others (De Gasparo and Van Belleghem, 2013). The *risk* then refers to the probability that psychosocial *disorders* should arise in and because of work.

Now that this definition has been clarified, one can consider psychosocial disorders at work as the symptoms of adverse impacts of work on the psychosocial dimension of everyday work, which relates to the psychical (e.g. motivation, engagement, subjectivity, values, etc.) and social (cooperation, mutual assistance, protection strategies) engagement of workers. This perspective is also followed by Clot (2010) when he suggests using the

---

* In France, psychosocial hazards are socially appointed, including by the government, such as psychosocial risks (*risques psychosociaux*). This ambiguous terminology, we believe, is inappropriate and should be discussed.

acronym PSR to designate psychological and social resources of people at work. These resources must be understood in order to be cultivated.

Indeed, this dimension should be recognized by ergonomics as one that contributes to structure activity (as a resource), and not just as a related feature (one that is considered to be poorly functional), viewed as some kind of failure of the work system, or as related only to the fields of psychology, such as, for example, occupational psychology (Clot and Kostulski, 2011) or psychodynamics (Dejours, 2012; Deranty, 2009).

Indeed, psychosocial disorders emerge precisely when the psychosocial dimension is not – or is no longer – recognized in its positive aspects by the work organization. Ergonomics needs to learn to make use of this discovery. It is also an opportunity for the discipline to recognize a dimension of work that it has forgotten. Yet, this recognition is by no means a revolution in activity-centred ergonomics, since it merely reminds us of the fact that personal mobilization is required for workers to deal with the discrepancy between prescribed work and real work that is present in every work situation (De Gasparo and Van Belleghem, 2013). Ergonomics must claim this dimension as its own.

From this point of view, if we define work as the real activity deployed by workers in order to achieve the goals they aim to achieve, one must consider that this activity is constructed based on the following:

- Set goals, prescribed procedures and means available to the worker (i.e. prescribed work)
- The real work situation, as it is in the moment where activity is carried out, which generates variability and unpredictability (real-world activity)
- The physiological, psychological and social dimensions involved in any activity, which allow:
  - A mobilization of the body in action: gestures, efforts, abilities, dexterity, etc.
  - A cognitive mobilization: representations, forms of reasoning, strategies, regulations, etc.
  - A psychical mobilization of the worker: motivation, engagement, competence, subjectivity, etc.
  - A collective mobilization: cooperation, mutual assistance, collective strategies, contributions to the rules of the trade, etc.

Here, the psychosocial dimension does indeed act as a dimension contributing to structure work in addition to the more classically considered physiological and cognitive dimensions. It makes it possible to deal with the events of real-world work by emphasizing the possibilities of action that are available to operators and by ascribing a subjective value to regulations that take place within work (allowing satisfaction related to a 'job well done'), the development of skills (allowing recognition), collective regulations (useful to cooperation), etc. It also relies on opportunities for debate (questioning, mutual assistance, attentive listening, etc.) and for thinking (keeping up a capacity for judgement, ensuring that action is consistent with one's own values) that are crucial to any activity. Being able to act, debate and think (Daniellou, 1998) are the conditions that are indispensable to the worker to face real-world situations. The psychosocial dimension structures them together by giving them a subjective consistency. However, this dimension is not a 'given'; it is constructed within and through activity itself.

The development of the psychosocial dimension: A driving force of activity

The psychosocial dimension of ordinary work is not set in stone. As are all of its other dimensions, it is in constant development. Indeed, it is through a confrontation with real-world events, which are always unpredictable and always complex, that this psychosocial dimension is called upon. It is also through this confrontation that workers find ways of 'working things out in spite of everything' – by inventing new ways of doing things, new ways of dealing with work situations, new ways of acting when faced with real-world constraints, new ways of cooperating, and by giving them meaning.

This renewed novelty is a part of the development of workers' activity. This is not just in the area of efficiency, which is characterized by an operational search for an answer to the requirements of production work within a theory of action, but also in the area of subjective engagement, corresponding to the involvement of oneself in work activity and in interactions with other people.

Therefore, any activity situation, whenever it implies a new mobilization of the subject, should be viewed as a situation of activity that is 'under development'. Similarly, any event – even it is to be considered a constraint – is an opportunity for workers to overcome, thus providing them with opportunities for operative and subjective development.

Every event processed in this way, whenever it is set within the workers' possibilities for acting, speaking out and thinking, is given new meaning. This new meaning is usually stronger than the involvement that was required to overcome the event. At a time where work shifted from 'a social definition where it was viewed as the rapid execution of gestures or elementary operations … to an approach where work can be considered as an intelligence and as a relevant conduct of events' (Zarifian, 1995, p. 7, our translation), one can easily understand the importance that the meaning ascribed to the events of work and how they are managed can have to the contemporary worker. The pride that workers can draw from having been able to deal collectively with an unforeseen and delicate situation is just as important as the specific know-how that they were able to construct at that time. Here, the subjective value of the activity is not unconnected from its operative value. The effectiveness of action does not just produce an effect – it also produces meaning.

Thus, the development of the psychosocial dimension strengthens the individual in the search for a *balance* when faced with the constraints of work – including one's own requirements (values, expectations, health etc.), the requirements of the activity (to contribute to the quality of production or of a service, etc.) and the requirements of the collective (to cooperate, to coordinate with each other, to support one another, etc.), including in situations where constraints are strong or particularly taxing, and rely very highly on engagement at work. In itself, maintaining this balance is a protection against hazards, whether psychosocial or otherwise.

Ergonomic activity analysis must strive to understand this *positive aspect* of the operators' mobilization in their own work – which organizes and gives structure to the relationship between *individuals*, their *activity* and *other people* on an everyday basis – and how it can develop. It must also understand what can hinder its development.

Obstacles to development viewed as professional hazards

Hazards occur when this balance is disrupted or broken. This happens when, in specific situations of overwork related to the variability present in work, the professional know-how of workers does not allow them to 'hold the various requirements of their work together'. The ability to achieve work goals may suffer as a result, as well as the meaning that workers gave to them – both as individuals and as a collective.

It is then that the first disorders appear – stress, tension with coworkers, exhaustion, etc. These may quickly develop, turning to disorders of more serious forms (interpersonal conflicts, psychopathological disorders, somatic diseases, etc.), with possible effects in the individual's personal life (addiction, marital problems, etc.) if these tensions persist or become more serious over time, with no possibility for elaboration, expression or resolution. Here, it is the obstacles to possibilities of acting, speaking out and thinking that are at the root of a situation of disrupted activity.

In these situations, the process of development of the psychosocial dimension is hindered, or stopped altogether. It no longer contributes to the construction of the resources necessary for dealing with future events. A vicious circle then forms, locking workers in situations of repeated failures in spite of increased efforts. In some cases, the failure to reach work goals is associated with the ungratifying task of having to include in scoreboards these costly regulation periods as 'nonproductive periods'. The psychosocial hazard – that is, the risk of seeing the appearance of a psychosocial disorder – becomes greater as the worker's mastery of the situation diminishes. The consequences of this affect both the employees involved – individually and collectively (in the form of psychosocial *disorders*) – and the results of work.

As when physical, chemical or biological factors are involved, hygienist approaches to work tend to consider that psychosocial hazards are related to factors that are external to activity itself (i.e. sources of danger or nuisance), which one should protect workers from by eliminating them at the source. On the contrary, we consider that hazards are intrinsically related to activity, and to the inability that it has of developing in some situations. Development protects workers from hazards. Obstacles to development generate them.

From there, the goal of prevention is not to protect employees from the impact of factors that are supposedly external to the activity, or even to preserve the existence of the psychosocial dimension of work within activity. It is to contribute to the development of activity in its various dimensions, including the psychosocial dimension, in order to allow employees to construct their own health. The design of enabling environments should contribute to this goal (Arnoud and Falzon, this volume; Falzon, this volume). A classical approach to prevention should be replaced here by a constructive approach to human activity. The development of this activity should be viewed as a strategic option in order to reduce hazards and improve work.

To achieve this, it is clearly necessary to gain in-depth knowledge of the processes that are at the root of this development of activity, in relation to the development of the subject, and of the social system.

The process of development at work

The gap between prescribed work and real work: A space invested by activity

Let us remind the reader of the starting point of our thinking. Work activity cannot be reduced to the simple execution of a *task*, as prescribed by the work organization imagined by F. Taylor. To understand what is classically meant by the 'gap between theoretical (or prescribed) work and real work', it is important not to attribute it too eagerly to the will of the worker alone. If an employee does not do exactly what is requested, it is not primarily because of unwillingness or lack of motivation, for example. Indeed, the relationship between a task and the worker charged with carrying it out should be considered in the context of the concrete and singular situation in which activity takes place. It is the here and now of the worker's current experience that is of interest to us, at the very moment in which action is undertaken.

When considering the real-world conditions of carrying out a task, the first experience of the worker is a mismatch between what was planned and the state of the world as it presents itself at the time of action. One might also speak of a 'resistance' (of the outside world and of the worker's own body) to the task, as a representation and as an anticipation constructed by the agent prescribing the work (who may, in some cases, be the worker himself). Things never go *exactly* as they had been planned. The 'reality of work' is therefore this space that is opened up by the existence of an always-irreducible gap between the theoretical representation of work, on the one hand, and the concrete and sensitive ways in which the state of the world presents itself to the worker, on the other hand. It is within this space, which is both inevitable and always specific, that the real-world activity of the worker takes place. It makes the worker something other than a simple executor. It makes him an acting subject in the world (Bourgeois and Hubault, this volume).

In the concrete process of work, 'the confrontation to the resistance of the real first gives birth to the subjective, affective experience of failure' (Dejours, cited by Deranty, 2010, p. 216) concerning forecasts, prior knowledge and previously constructed procedures. The gap between prescribed work and real work indicates that before doing what has been asked of them, workers are first confronted with a problem or an unforeseen event that prescription (e.g. tasks, allocated means, directly available information) is not able to solve completely. It is to cope with this resistance of the real world to prescription contents that workers must engage themselves personally, in order to discover and invent an original outcome for action, making it possible to reach the desired goal. This mobilization involves workers in all the dimensions of their being: efforts of the body, sensitivity, technical know-how, ability, ingeniousness, the knowledge acquired through experience and the more formal knowledge of symbolic systems that are specific to a profession. This mobilization is motivated by many factors.

In this way of committing oneself 'body and soul' in a real-world situation of activity, subjects involve part of themselves in the scene of work. They awaken their bodily, cognitive and affective sensitivity to better understand what is going on. They recover knowledge acquired previously, in other circumstances. They apply skills to test new solutions. They request the assistance of other people who are able to help them. They take the risk of drifting away from formal prescriptions in order to achieve their goals. Here, there is a total commitment of the worker, who makes available for work some components of oneself that go far beyond what is requested and expected in order to respond to events in the real-life situation of work. Whereas a task may forecast relying on a specific resource that is specific to a given individual (e.g. physical strength, ability to carry out an operation, to solve a specific problem), real activity requires the mobilization of all components of this individual.

Activity can then be defined as a *global mobilization* of the acting subject, aiming, at his behest, to find original and effective ways to do things in the face of the reality of situations.

Subject and activity: A joint development

In the sense that 'what I am doing' (my action, its results, the quality I expect in them) mobilizes part of 'who I am' (my body, my knowledge and also my initiative), one can easily understand how subjective involvement in work activity is a major issue for the mental and psychical health of workers.

Yet, confrontation with the real world is always unique and singular. Every time, the situation is new and different. This implies that mobilization is never a mere replication of conducts or solutions identified in the past. The search for a suitable response, aiming for a certain degree of quality in work, involves a form of creativity and invention that is related to a learning process. This process is not limited to the acquisition of formal knowledge, but may potentially extend to all aspects of the existence of the acting subject.

We propose the term *sedimentation* to refer to this added value created in the act of work, that the subject derives from work. This sedimentation may take on various forms, and should not be understood as the simple superimposition of successive layers. Sedimentation can enrich, modify and disrupt the way an individual relates to himself or herself over time. It is a fertile ground from which any future mobilization will stem (Figure 3.1).

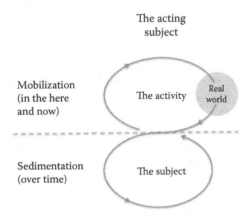

Figure 3.1 The process of mobilization and sedimentation in the acting subject.

Therefore, the process of development is twofold (see Figure 3.1). It involves, simultaneously, a process of mobilization of the subject in the activity and a process of sedimentation of activity within the subject (Delgoulet and Vidal-Gomel, this volume). 'What I do is what I am'. Within this double loop connecting the subject to the real situation where activity is carried out, the dynamic process of mobilization and sedimentation occurs. Development focuses both on the drivers of the activity of the acting subject and on the various domains of its existence.

However, the development of the psychosocial dimension does not impact just the subject taken individually. By definition, it also extends to the social system.

The mirror dynamics of the acting subject and the social system

The encounter between the subject and reality is not made in isolation. Beyond divisions of work imposed by the organization, the activity of an individual is always connected with other individuals, via the social interactions required for their coordination (Boissières and de Terrsac, 2002). These interactions rely on sharing a framework of social norms, which are known and acknowledged by all (see Caroly and Barcellini, this volume). These social norms define goals, task attributions, instructions, rules and regulations to comply with, deadlines to reach, evaluation criteria, modes of control and systems to use. They regulate the circulation of information, the means of communication and cooperation, the management of tension within the collective, the means to capitalize on experience and the technical means available. They set markers for what should be done and give criteria for engaging in an action and understanding its scope.

This set constitutes a 'common good' and provides resources to deal with unforeseen events, tensions and the events that occur. It should not be a source of constraints (de Terssac and Gaillard, 2009). Beyond the fact that it has a prescriptive character regarding the means to be used and the processes to be applied, it also conveys, in a more or less explicit manner, values and preconceptions regarding the work to be done and the trade.

What the reader needs to understand here is that these social norms were not elaborated 'in some place other' than in collective activity. They are the product of the sedimentation of the interactions between all of the agents within the system in time, whether these agents are 'mere operators' or 'major decision-makers'. Social norms capitalize organizational decisions. They show agreement and disagreement regarding how the work should be done. They are subjected to the filter of the reality of work situations and collective experience. They have successfully passed the trial of sharing – or indeed, of confrontation – of the knowledge and

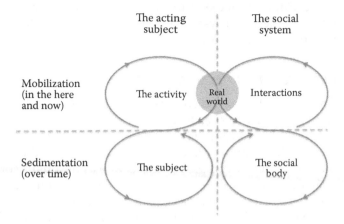

Figure 3.2 The process of mobilization and sedimentation connecting the subject with the social body.

expectations of decision-makers and those who act upon them in practice. Finally, they contribute to the concrete definition of organizational rules.

In other words, the organization of work is always a product of a work of organization that structures the norms of the social body and allows it to act collectively, but also to recognize itself as an entity bearing shared values, legitimacies, rules of authority and delegation, principles for action and decision, etc. Just as the subject is the crucible of the sedimentations of activity, the social body catalyzes the sedimentations of norms elaborated within collective interactions (see Figure 3.2).

In return, the social norms elaborated within the social body guide action and constitute resources for it. They constitute a fertile ground allowing the mobilization of the work collective in order to cope with the reality of work situations in the here and now of activity. Thus, they contribute to connecting the subject with the social body, not in an abstract connection of one with the other, but via activity, at both the individual and the collective level, that is implemented to cope with real-world events toward a shared goal. Thus, they contribute to structuring the social dimension of work, in an intimate relationship with its psychical dimension.

Once again, this process of mobilization and sedimentation is not set in stone. It is the result of a constantly renewed dynamic process that occurs within the time of action, contributing to the joint development of social interactions (in terms of collective effectiveness, cooperation, coordination, etc.) and the social body (in terms of rules, shared values, collective moral principles, etc.).

Thus, the development of the acting subject occurs at the same time as the development of the social system as a whole. The driving force of the psychosocial dimension of work is set in motion.

Development as an end goal for ergonomics

The occurrence of psychosocial disorders suggests a block in this double process of development, affecting both the social body (tensions between coworkers, aggressive behaviour against users or customers, conflicts of rules, questioning the legitimacy of authority figures, etc.) and the health of workers (stress, anxiety, depression, etc.), as well as the quality of work (failure to manage unforeseen events, system malfunctions, drops in quality metrics, failure to provide users with a service, etc.).

The development model makes it possible to view in another light the emergence of disorders in organizations. Classical approaches to psychosocial hazards (e.g. safety, medical and insurance based) are guided by the quest for attribution (of a cause, etiology or blame). Following this search, the next logical action is to remove the risk factor identified in this way, suggesting that there are superfluous elements in work that should be dismissed. And yet, it is precisely the opposite that occurs. Activity is suffering because it is drained, and amputated of some of the resources that allow it to develop in order to ensure that, in real-world situations, quality of work, quality of social relations and individual commitment all occur. It suffers from a shortage, not from a surplus that should be reduced.

Hence, a violent reaction on the part of a user in a public service space (e.g. a sales point, information desk, administrative service, etc.) is often interpreted as the result of latent aggressiveness being on the rise in the population. This interpretation often leads to the implementation of systems intended to protect agents, e.g. antiassault windows. We might, however, propose a different analysis of these situations, in terms of hindrances to development. A violent situation emerges, for example, from an increasing mismatch between the commercial promise made to users by the company and the means that are effectively available to the agent to this promise when carrying out a transaction. At this point, the development of cooperation between the user and agent, aiming to co-construct a high-quality service for both parties, is hindered, in spite of the will of the first party to 'do a good job' and of the second party's benevolence. One can then consider different actions than physical protection. These must aim to match the commercial promise (or service provided) with the means effectively made available to agents, in order to achieve a constructive development of the service relationship.

Therefore, ergonomic interventions must aim to work situations analysis while considering the issues related to the development of the psychosocial dimension. It must ensure that development processes occur for each of the situation's protagonists, both in their operative component (mobilization) and in their subjective or social component (sedimentation). Last but not least, it must move beyond a posture of analysis to set off, with all of the stakeholders involved, a process aiming to transform work situations. This transformation must explicitly aim to spark a debate about work and the conditions in which it is done, by setting up spaces for regulation and discussion (Detchessahar, 2011).

Ergonomics must develop new tools to deal with this debate. It is relatively new to the field, insofar as it explicitly aims to develop the psychosocial dimension of work. Simulating work (Barcellini et al., this volume) – and more precisely, simulating work organizations (Barcellini and Van Belleghem, 2014) – aims to define acceptable rules for activity, and should be able to contribute to this effort. The goal is to implement participatory systems to design or transform work situations in which workers are requested to 'play out' their own work by using an adequate simulation device. This simulation of activity makes it possible to explore and assess possible prescription scenarios. It allows the various stakeholders involved in the transformation (e.g. operators, decision-makers, prescribers, staff representatives, etc.) to engage in a debate regarding the rules and the means of work, and the quality that is expected of this work. It must make it possible, on this occasion, to assess the operative and subjective potential of the choices made (e.g. the a priori interest of the tasks attributed to the workers, their relevance to the profession, the relevance of resource allocation, possible mismatches between expectations of the organization and those of its employees, etc.). Finally, simulation must allow stakeholders to give meaning to the choices they make, when they are the product of trade-offs. By taking part in their elaboration, operators are able to give these rules a meaning even before they are implemented.

Simulating an activity is itself an activity, thanks to the active participation of operators; it also sets off a process of mobilization and sedimentation. The choices discussed and made during the simulation (mobilization) produce meaning for operators, and contribute to the elaboration of new norms for the social body (sedimentation). In other words, the simulation of work is an opportunity to 'play out', beforehand and on a small scale, the development process that may occur following the transformation of the situation. When this process is carried out based on several prescriptive hypotheses, it can be speeded up, by allowing several paths of development to be explored and the most relevant path to be selected.

This methodological approach must be consolidated with further methods. Ergonomics should be able to open and support new spaces for discussion and debate about work (participatory approaches, experience feedback, etc.) that make it possible to support decision-making with a rich view of activity, whether the decisions involved are of a technical, organizational or social nature. These spaces should make it possible to respond to Detchessahar's suggestion (2011, pp. 100–101) to steer organizations toward 'discussion engineering', aiming, in this case, to 'organize the work of organizing', whose explicit goal would be to ensure the socio-organizational balance of the company and the psychical health of its employees.

This path carries a meaningful prospect for ergonomics: to turn the development of the psychosocial dimension of work into one of its goals.

References

Barcellini, F., and Van Belleghem, L. (2014). Organizational simulation: issues for ergonomics and for teaching of ergonomics' action. In O. Broberg et al. (Eds.), *Proceedings of Human Factors in Organizational Design and Management.* XI Nordic Ergonomics Society Annual Conference, 46. Santa Monica, CA: IEA Press.

Boissières, I., and de Terssac, G. (2002). Organizational conflicts in safety interventions. In B. Wilpert and B. Fahlbruch (Eds.), *System safety: challenges and pitfalls of intervention* (pp. 119–131). Oxford: Pergamon.

Clot, Y. (2010). *Le travail à cœur.* Paris: La Découverte.

Clot, Y., and Kostulski, K. (2011). Intervening for transforming: the horizon of action in the clinic of activity. *Theory and Psychology,* 21(5), 681–696.

Daniellou, F. (1998). Une contribution au nécessaire recensement des "Repères pour s'affronter aux TMS". In F. Bourgeois (Coord.), *TMS et évolution des conditions de travail: les actes du séminaire, Paris 1998* (pp. 35–46). Lyon: ANACT.

De Gasparo, S., and Van Belleghem, L. (2013). L'ergonomie face aux nouveaux troubles du travail: le retour du sujet dans l'intervention. In F. Hubault (Coord.), *Persistance et évolutions: les nouveaux contours de l'ergonomie.* Toulouse: Octarès.

de Terssac, G., and Gaillard, I. (2009). Règle et sécurité: partir des pratiques pour définir les règles? In G. de Terssac, I. Boissières, and I. Gaillard (pp. 13–34), *La sécurité en action.* Toulouse: Octarès.

Dejours, C. (2012). From psychopathology to psychodynamics of work. In N. Smith and J.-P. Deranty (Eds.), *New philosophies of labour: work and the social bond* (pp. 209–250). Leiden, The Netherlands: Brill.

Deranty, J.-P. (2009). What is work? Key insights from the psychodynamics of work. *Thesis Eleven,* 98(1), 69–87.

Deranty, J.-P. (2010). Work as transcendental experience: implications of Dejours' psycho-dynamics for contemporary social theory and philosophy. *Critical Horizons: A Journal of Philosophy and Social Theory,* 11(2), 181–220.

Detchessahar, M. (2011). Santé au travail. Quand le management n'est pas le problème, mais la solution…. *Revue Française de Gestion,* 5(214), 89–105.

Herbig, A., and Palumbo, F. (1994). Karoshi: salaryman sudden death syndrome. *Journal of Managerial Psychology,* 9(7), 11–16.

IEA, International Ergonomics Association. (2006). *50th anniversary booklet.* Santa Monica, CA: IEA Press.

Johnstone, R., Quinlan, M., and McNamara, M. (2011). OHS inspectors and psychosocial risk factors: evidence from Australia. *Safety Science,* 49(4), 547–557.

Kortum, E., Leka, S., and Cox, T. (2011). Perceptions of psychosocial hazards, work-related stress and workplace priority risks in developing countries. *Journal of Occupational Health*, 53(2), 144–155.

Leka, S., Jain, A., Iavicoli, S., Vartia, M., and Ertel, M. (2011). The role of policy for the management of psychosocial risks at the workplace in the European Union. *Safety Science*, 49(4), 558–564.

Lippel, K., and Quinlan, M. (2011). Regulation of psychosocial risk factors at work: an international overview. *Safety Science*, 49(4), 543–546.

NDCVK, National Defense Council for Victims of Karoshi. (1990). *Karoshi: when the "corporate warrior" dies*. Tokyo: Mado-Sha.

Salher, B., Berthet, M., Douillet, P., and Mary-Cheray, I. (2007). *Prévenir le stress et les risques psychosociaux au travail*. Lyon: ANACT.

Shain, M. (2009). Psychological safety at work: emergence of a corporate and social agenda in Canada. *International Journal of Mental Health Promotion*, 11(3), 42–48.

Van Belleghem, L., and De Gasparo, S. (2014). Entry: trouble psycho-social. In P. Zawieja and F. Guarnieri (Coords.), *Dictionnaire encyclopédique des risques psycho-sociaux* (pp. 807–808). Paris: Editions du Seuil.

Zarifian, P. (1995). *Le travail et l'événement, essai sociologique sur le travail industriel à l'époque actuelle*. Paris: L'Harmattan.

chapter four

From the adaptation of movement to the development of gesture

Yannick Lémonie and Karine Chassaing

Contents

Although the issue of gestures at work is not a new one, evolutions in population demographics and in work continue to give it some prominence. The goal, newly reaffirmed, of standardization of work, where gestures are the focus of increasingly strong prescriptions, or the increase of musculoskeletal disorders (MSDs), makes it necessary to revisit the models and action strategies used by ergonomists to tackle the issue of gestures at work in their transformative actions.

The goal of constructive ergonomics lies in the design of enabling environments for work. The characteristics of such environments are as follows: nondetrimental, universal and allow development and learning (Pavageau et al., 2007). Following this, interventions on professional gestures must allow the identification of constraints that cause the emergence of strategies that are detrimental to operators, in order to allow the preservation of gestural variability as a resource constructed by operators (1) to be effective, (2) to protect themselves and (3) to open up reflective spaces that allow these gestures to be discussed. These three levels of action are essential.

Thus, taking the opposite stance from prescriptive and normative models, action on gestures at work can only take place by recognizing its intentional, creative and reflective dimensions. In other words, the goal is not so much to prescribe a 'best gesture' as to provide resources that allow operators to develop their own gestures in order to be effective at work, protect themselves and build themselves up. This implies viewing a gesture as a construction, and not as a mere execution, viewing the variability of gestures as a resource for the organization of work and not as an obstacle, and viewing reflectivity as the driving mechanism in gestural development.

We will structure our argument in three parts. In the first section, we will see that an understanding of the mechanisms whereby motor acts are produced and controlled makes it possible to act upon the constraints of the task and environment, so that operators might be able to produce gestures that are nondetrimental to their health. Although this first level of action is necessary, it is not sufficient, since it tends to reduce the gesture merely to its execution. By taking seriously the active part that the operator plays in elaborating the gestural solution, we will argue in the second section that gestural variability is a resource that makes it possible for operators to regulate their activity and preserve themselves. In this sense, an ergonomic intervention must also allow the preservation of this resource for operators. However, this does not imply anything about the possibilities for the development of gesture. In the third section, therefore, we will argue that the development of gesture can only take place by opening up a reflective space allowing operators to put this gestural variability to work.

Acting on the constraints that affect movement

Understanding the impact of work situations in terms of constraints is not new in ergonomics. Indeed, one of the first directions of work for ergonomics is to ensure that the work environment does not constrain the strategies of operators in ways that are detrimental to their health or their effectiveness. Models, mostly biomechanical in nature, that focus on movement illustrate this vision of the work of ergonomists, by focusing on solutions related to the geometry of workstations.

However, biomechanical analysis is not sufficient to consider all of the solutions that ergonomists can propose in terms of workstation design. Thus, Aptel and Vezina (2008) insist on 'the need to take into account the role of motor commands to understand the impact on operators'. Movement is no longer viewed solely in its effective capacity, but as the product of a 'complex and integrative system that is of a psycho-cognitivo-sensorimotor nature' (our translation). Following this view, the explanations provided in the field of neuroscience to explain the production and control of voluntary movement have rested exclusively on a computational approach, up until the past decade. This theoretical orientation postulates the existence of a central control of voluntary movement by the nervous system. This sets the organization of all of the body's degrees of freedom, which are stored in the central nervous system (e.g. Schmidt, 1975).

However, this approach has been questioned by the sudden entrance of models derived from the analysis of dynamic nonlinear systems (e.g. Kelso, 1995). Such models postulate, based on the ideas of Bernstein (1967), that the central nervous system is incapable of controlling all the degrees of freedom of the body when it is producing a complex movement. Following this perspective, a complex movement is viewed as emerging from a network of constraints. Motor commands are not centralized within the nervous system, but reside within the dynamics of interaction between the individual and his or her physical environment. Depending on the level of constraints, many preferential patterns may emerge. For example, when imposing a low speed of locomotion to a subject on a treadmill, this subject may adopt one of two motor solutions: walking or trotting along. When the level of constraint rises, motor solutions tend to become more restrictive. Thus, when a greater speed is imposed, the preferential pattern that emerges is running. The subject may only adopt a different motor pattern, such as walking, at a very high cost in energy (e.g. Brisswalter and Mottet, 1996).

This example leads us to argue that one of the first levels of action for ergonomists is to identify and act upon the constraints that affect movement, causing the emergence of types of coordination that are potentially ineffective or detrimental. It is thus possible to identify levels of constraints that are likely to reduce the variability of possible types of coordination and to augment the harmful effects of repetition of movement. Indeed, the rhythm imposed by a task, as well as the unforeseen events that may occur in any work situation, coupled with workstation design, is likely, beyond a certain threshold, to limit the possibilities of coordination adopted by the operators.

An example from the food processing industry can be used to illustrate this point. The pace of a processing line for deveining *foie gras* (removing the veins from the liver) is set by considering the time required to carry out operations on the product, but considering all of the products as being identical. Yet, products such as animal livers are variable: some are tougher than others, larger or smaller, etc. Thus, the operations that need to be carried out do not require the same time for each liver. Operators remove the veins using a knife, holding it midway between the handle and the blade. All of them do not wear gloves, so as to reduce the sliding of veins between their thumb and the blade of the knife. While they are removing the veins using the knife with one hand, the other hand is holding the liver fast so that it does not tear and that too much produce does not come away with the vein. Calculating the rhythm of the line sets the spacing between the livers, and operators working along the line do not have enough space to regulate the variability of the product. Furthermore, task instructions prohibit lagging, i.e. falling behind and moving down the line. Risks emerge notably in terms of MSDs, along with phenomena such as self-acceleration, pressures, and contractions to hold the pace, a reduction in gestural variability, and a feeling of being unable to do 'quality work':

> 'We do quantity, not quality'.
> 'Quickly done is badly done'.
> 'It is not what I do that I don't like. It's the conditions which I do it in'.

Operators are not free to choose their gestures in production time. They cannot cope in real time with the unforeseen events of production.

Some constraints other than an imposed rhythm are liable to explain the adoption of inefficient or detrimental forms of coordination by the operator. Newell (1986) pointed out three types of constraints: those related to the task, to the environment and to the organism. To draw upon an example given by Bril (2012), multiple constraints are at work that interact with one another and organize movement when bearing a load: the nature of the ground, the weight of the load and the distance to travel, but also the aspects related to the potential of the organism (in physiological, cognitive, affective terms, etc.).

Dynamic models that combine cognitive and biomechanical approaches are useful when acting on the constraints affecting the voluntary movement of operators. Yet, they remain quite reductive, since they do not account for the active part that is played by operators when searching for and producing effective and efficient motor solutions to respond to task requirements. Hence, these models tend to reduce gesture to movement.

Although a movement is the observable part of the gesture, it can and should by no means be reduced to it. On the contrary, gestures are complex by nature – that is, they cannot be reduced to any one of their many dimensions: biomechanical, psychological, social, contextual or cultural. From this point of view, the biomechanical analysis of gestures is reductive, since gestures, as complex entities, imply a holistic rather than an analytical approach. Furthermore, gestures cannot be set apart from a cultural history, nor from the history of the working environment and of the transformation of work situations. Thus, the concept of gesture is close to the concept of bodily technique, as defined by Vigarello (1988), following from the works of Marcel Mauss: 'the transmissible physical means that are deemed most adequate to reach a goal in a given situation' (our translation).

This minimal definition makes it possible to understand that a gesture is a motor solution that the operator has identified to be both effective and efficient to achieve his or her set goals. The goal of the work of the ergonomist is, from this point of view, to design work situations that allow operators to implement gestural solutions that are suited to the requirements of the current situation, as well as to the requirements of the operator in question.

Gestural variability: Opening up the space of possible solutions

A gesture is a trade-off, a solution constructed by the operator at a given moment to respond to the requirements of a task. The task can then be viewed as a problem space (Durand, 1993). However, this problem space is never frozen in the context of a work situation. It possesses its own dynamics, which are the product of the variability of situations encountered by operators: the variability of products and constraints, the variability of environmental conditions and the variability in the operator, as shown in recent research on working times (e.g. Barthe and Quéinnec, 2005; Toupin et al., this volume). Because of this, it is reasonable to believe that gestural variability is a resource constructed by operators to fit to the specific dynamics of task constraints, the environment and their own state.

Furthermore, gestural variability allows operators to protect themselves from MSDs (e.g. Madeleine, 2010). By adopting a mode of operation that allows recuperation of the tissues solicited during another strategy, this gestural variability allows some form of repetition with no monotony. Not repeating gestures in identical forms, and allowing gestures to vary, allows the worker to solicit parts of his body in different ways. It also allows breaking monotony, to create gestural variants, to search for the gestural solution that is best suited to oneself and to the situation at hand. In the event of these creative processes being hindered, the gesture is cut off. It is the result of a hindered activity that 'locks the activity into identical repetitions, involving the subject in compulsive activities where the motor aspect is no longer handled through automatisms, but through synkinesia. The latter refers to a system of movements that can only be executed together, and always in the same way' (Clot and Fernandez, 2005, p. 74, our translation).

As we can see, gestural variability is a resource constructed by operators to cope with the unforeseen elements and the variability of work situations. It is an indicator of their skill and accounts for the active part that they take from and commit in their work (Bourgeois and Hubault, this volume). From there, it becomes necessary to be mindful of this variability in order to support the effectiveness and efficiency of actions targeted by gestures. They constitute a resource for system performance. The design of work systems implies defining leeway for gestures in real-time production and in learning.

Let us illustrate this idea with an example from a company in the automotive sector. In this company, which has implemented measures for work standardization, managers are notably tasked with defining gestural prescriptions. These prescriptions are given in 'operation sheets' (Chassaing, 2010). These operation sheets present only one way of doing things, and there is no possible choice for the operator, thus leaving little room for the variability of gestures. Here is an example of a prescription:

> Using both hands, take the lining (i.e. the piece forming the inside casing of a car) from the TM (a large container). Placing the left hand in the oblong openwork, next to the wheel's passage space, the right hand should be in the central part of the rear side panel. When leaving the TM, rotate the piece 90 degrees right.

This prescription restricts operators as they are looking to balance the piece, to control it, to spread their arms depending on their height. Because of this, they adopt modes of operation that are widely different from what is recommended by the prescription. This prescription becomes a constraint for operators, who position their hands differently on the lining, so as to handle it differently from what is prescribed. These gestural solutions allow operators to achieve their goals related to safety, comfort, muscular fatigue and effectiveness, which are not taken into account in the instructions. On the contrary, prescriptions should not constrain modes of operation, but allow and indeed encourage operators to construct a gestural solution. Here is an example of a prescription that offers potential margins of manoeuvre for operators to construct gestural solutions:

> Place the lining in the assembly, simultaneously matching the top left guide of the assembly with the top cylindrical hole of the lining, and matching the central guide of the assembly with the central oblong hole of the lining.

Here, the prescriptor only suggests guides regarding the results of action. The position of the hands is not mentioned, and this operation therefore leaves operators with some freedom regarding how to do the job.

However, the operations present in the instruction form a single unit. They are dependent upon one another. Some of them leave some leeway, while others do not. The execution of an action that is described with potential leeway, leaving some space for various possible modes of operation, can be constrained by the preceding operation that may or may not be described in greater detail. Such is the case with the example cited above. Indeed, the prescription regarding how to handle the lining implies a certain position of the hands. This prescription precedes the one concerning the placing of the lining in the assembly, which, conversely, provides some leeway regarding how the lining should be placed. This potential leeway is reduced by the rigidity of prescriptions concerning the preceding stage.

The question that remains at this point is how not to inadvertently reinforce restrictions to the diversity of modes of operation, and therefore:

- To provide prescriptors with the means, in terms of time and training, to understand the causes of diversity in the activity of operators occupying the same station, in order to take into account, at least in part and in a relevant manner, this diversity.
- To favour 'justified' prescriptions – i.e. those for which there are clearly established stakes in terms of quality and safety that can in turn be explained to operators.
- Following this twofold goal, and in general, not to encourage the production of instructions that are detailed step by step, or compliance with all of these points without referring to their importance. Some form of prioritization seems justified in the use of these instructions.

The goal, in terms of designing prescriptions for work procedures, is to readily include the perspective of the diversity of gestures, both within individuals and between individuals. Following this view, if there is a need for a physical support to define what operations should be carried out, this support might look rather like a 'guide to the activity', which might serve, on the one hand, to propose operations while describing their merits and drawbacks, notably to be used by operators in training, and on the other hand, to collect and confront gestural variations constructed by each worker to support professional debates' focus on gestures at work and mutualization of practices (Vézina et al., 1999). The prescription may then become a source of reflection on practice. The analysis of the activity of experienced workers, carried out by an ergonomist, then becomes a tool to identify and extract incorporated knowledge, so as to contribute to the design of prescriptions that are mindful of the variability and the diversity of people and work situations.

Similarly, training should encourage the construction of a gestural solution that takes into account these various forms of variability. One can only be sceptical, at this point, regarding training programs that focus on the 'one best gesture' and 'one best posture', which tend to reduce the scope of the solutions implemented by operators and to decontextualize the solutions constructed in a specific context. On the contrary, these should be designed based on the analysis of gestural variability and on the identification of gestural know-how, in order to allow trainees to construct their own gestural solutions based on those of more experienced operators.

Here is an example. In a company from the food processing industry, the design of a duck cutting line provides some room for manoeuvre, notably to organize the training of new employees on the line without disrupting production (Coutarel et al., 2003). An intervention carried out 5 years after the implementation of this line made it possible to study in greater detail how operators are trained to the gestures of cutting (Dugué et al., 2010). An experienced worker, who guides the operator in real time during the production phase, supervises this training program. The trainer is acting in a doubloon. He demonstrates the gesture, divides the action into more elementary operations so that the apprentice can learn step by step, and examines and retains what the apprentice is not doing. This is a contextual training to gestures, where cutting gestures are apprehended in all of their complexity: in terms of effectiveness, efficiency and health. The trainer demonstrates gestures, accompanies the gesture of the apprentice, and emphasizes the quality of the cut product and the strength that should be exerted on the knife. A few hints are given, allowing the apprentice to strain less and to do high-quality work while holding the pace. For example, in order to cut the leg of a duck, part of the cutting is done 'blind', in the sense that the operator cannot see the joint. The trainer explains, by demonstrating and guiding the knife into the joint, that this joint is shaped like an S, and that one should follow this shape to cut the leg well (and not damage the bone), not to strain on the tough parts of the bone, and not to waste time by jamming the knife into these tough parts. The training places an emphasis on the sensation of using the knife when following the S-shape, and when the blade encounters obstacles when reaching the tougher parts. Some other knowledge is also imparted, regarding topics such as the angle of the blade against the flesh, the depth the knife should reach in the flesh, etc. The diversity of sensory information that operators rely on can be used as an input for training programs. This sensory information is essential for carrying out a gesture. And often, this information and its role in the gesture are underestimated – as evidenced by training programs that are mostly based on a purely biomechanical view of gestures. Furthermore, the learner is immediately confronted with the variability of the situation. The knowledge underlying the gesture of cutting a leg, such as making an S-shape, is manipulated in a variety of contexts, that is, in real production time and on a variety of ducks. This diversity of context becomes a source of reflection in the constitution of new modes of operation.

Although gestural variability is essential, it does not presume that it is possible to develop the chosen gestures and modes of operation. In the design of workstations, prescriptions, training systems, etc., it is not enough to allow for and foresee gestural variability for operators to develop new gestures. It is also important to open up reflective spaces, spaces and times to develop gestures.

Opening up reflective spaces for the development of gestures

Reflectivity plays an irreplaceable part in the development of gestures at work. This is because gestures are inseparably productive and constructive (Delgoulet and Vidal-Gomel, this volume; Rabardel and Samurçay, 2001). Obviously, the gesture makes it possible to perform a task and achieve a productive goal (the productive dimension). Simultaneously, however, it makes it possible to construct one's own experience (the constructive dimension). This makes it possible to understand how knowhow and experience allow operators to protect themselves. The issue that arises for ergonomists is therefore: how can one design work environments that allow the construction of this experience and the development of gestures? Reflectivity is the driving engine in this development, and the goal is to include leeway in work systems, allowing operators to become reflective practitioners (Schön, 1983), not simple underlings.

From a cognitive viewpoint, gestural expertise manifests itself through the incorporated (Leplat, 1995) and largely implicit and tacit (Polanyi, 1969) character of the knowledge underlying gestures. Most scientific works on the subject agree that the development of expertise in gestural production is accompanied by a less cognitive effort. For example, in Rasmussen's (1983) Skills, Rules, Knowledge (SRK) model, the three levels of expertise are characterized by a specific level of internalization/externalization. Within this framework, the sensorimotor level of skills is the level that is the most internalized and the one that can least easily be elicited.

In acknowledging this fact, several questions emerge: What are the lessons that operators can draw from their own experience, if the most internalized dimension of that experience is largely implicit and incorporated – and therefore is the one that is least accessible to verbalization? How can ergonomists act on situations to facilitate externalization, as well as reflection on the incorporated and tacit dimensions of gestural experience?

*From elicitation to reflection on action: The ergonomic
intervention as a means to create a reflective space*

To draw from one's experience implies that the operator undertakes a
reflection on the gestures that he has made use of – in other words, a metafunctional activity (Falzon, this volume). This reflection implies distancing
himself from his work. The issue here is to make it possible for operators to elicit the procedural knowledge used in the execution of gestural
patterns. It is this shift, from the implicit knowledge incorporated in a
gesture to the elicitation of this knowledge in verbal form, that needs to be
explored in order to derive prospects of action for ergonomists. The use of
methodological frameworks for a posteriori verbalization clearly relates to
this logic: auto- and allo-confrontation interviews (e.g. Mollo and Falzon,
2004), elicitation interviews (Vermersch, 1999), or alternately, in resitu subjective interviews (Rix and Biache, 2004). Although we will not describe
the detailed methodology behind these techniques, one can note that the
part played by the ergonomist consists in supporting the description of
the experience of a gesture mobilized in a singular situation. Hence, these
methodological devices are liable to create the conditions for externalization, and for updating the tacit knowledge that underlies the gestures
of operators.

Two interrelated stages can be outlined when eliciting the procedural
knowledge mobilized in gestures (Six-Touchard and Falzon, this volume):
a first stage of externalization, i.e. 'putting into words', which can be
termed elicitation, and a second stage from which it is possible to engage
in a reflection about action.

However, it is not always necessary to create a separate 'space-time'
for operators to be able to elicit the tacit knowledge incorporated in their
work. For example, Fillietaz (2012) has shown that during activity analysis, operators are likely to show, to demonstrate, to put into words some
of their knowledge in the situated context of their work by adopting a
reflective stance. This 'putting into words', which Fillietaz calls 'situated
elicitation', constitutes, in fact, a set of true opportunities for professional
development. The ergonomic intervention is liable to play a significant
role in the elicitation of gestures, in the sense that operators aim to describe
and help understand their ways of doing things – adopting a reflective
stance de facto.

The role played by the sources of variability encountered in work

Beyond the creation of a separate space-time and the effects of the
ergonomic intervention, we can also identify further effects that are
liable to provoke a realization and a reflective stance on the part of
operators. Between the goal and the result of an action – which are both

conscious – one can realize the means of one's own action, when encountering failure or when, for one reason or another, the subject aims to know the modes of operation adopted and their relationship with the results of action. Thus, there is realization when the operator is faced with occupational obstacles. From this point of view, the role of the ergonomist cannot be to erase all the difficulties from work. Instead, it will be to set up work situations so that they incorporate, in a central position, 'enabling constraints' (Davis and Sumara, 2007), i.e. constraints that allow the development of gestures at work (Delgoulet and Vidal-Gomel, this volume). In this situation, the variability of work situations can be considered a means for realization and for the involvement of the operator's reflectivity. To draw once again from the example of the deveining line in the production of *foie gras*, the variability of livers can constitute a source of reflectivity for the operator to elaborate an effective and efficient gesture – provided, in particular, that in the design of the production line, the space-time relationship allows such gestural regulations. In practice, the operator can activate a reflective activity through repetition, in order to identify parent situations, variations, constants and variables in order to modify, develop and adjust the gesture.

The work collective at the service of reflectivity: Transmitting, capitalizing and putting gestures to work

A final dimension of gestures is the fact that they are rooted in a professional culture, which some authors have called a professional genre (Clot and Faïta, 2000). Any trade comprises an inventory of know-how, iconic techniques and forms of recognition of professional competence. The collective is a bearer of this shared culture of the trade, and is also a crucial resource for development. Although gestures can only be acquired through personal experience, one seldom learns alone. The acquisition of experience is facilitated by those who have already acquired this experience, and in this sense the collective constitutes a resource in the construction of a gesture at work (Sigaut, 2009). In the learning of a gesture, the collective passes on to its new members a shared experience of the trade. From this point of view, debating the gestures of work is likely to become a 'psychological instrument' at the service of the members of the collective. By drawing upon obstacles to the activity and personal inventiveness, these debating practices bear new prospects for realization (Simonet, 2011).

Discussion spaces focusing on action and ways of doing work are needed. This entails that work situations should be designed in order to allow the physical presence of a colleague at the workstation, and that the instructions provided should also allow mutual aid between operators. This will allow the creation of a discussion space focusing on action and on ways of doing work. Such discussion spaces are intended to support potential debates between peers focusing on gestures, as we have seen in the case of the duck cutting line. The presence of a colleague at the workstation is all the more valuable for debating gestures that cannot be readily 'put into words'. Operators can demonstrate gestures, and perform them while commenting them, in order to support the elicitation of tacit gestural knowledge.

As Pastré (1997) has pointed out, some individuals will be able to make use of their errors, failures and successes, whereas others will repeat the same behaviours over and over again without being able to adapt them. To account for this difference between people, the author mentions the idea of 'taking advantage from' past experiences. This equates the concept of reflective practice, which is a crucial condition for conceptualization. He distinguishes two kinds of experience based on what they produce: an 'experience that locks the subject into the automation of his/her conducts, and an experience that opens, even in a limited way, prospects that go beyond mere experience' (Pastré, 1997, p. 90, our translation). Therefore, according to the author, experience is constructed based on 'the ability of a subject to go back on what has been lived in the past, in order to analyze and reconstruct the know-how at another cognitive level. By making use of the past, the subject extends his capacity for anticipation, opening up more broadly to the future and to the field of possible futures' (Pastré, 1997, p. 91, our translation). The ability of individuals to go back on what has been lived in the past – an essential condition to the development of gestures – is a crucial aspect of ergonomists' interventions. The collective plays a key part in these reflective analyses, and therefore in the development of gestures.

Conclusion

An understanding of mechanisms involved in the production and control of voluntary movement can usefully complement biomechanical analyses. These analyses make it possible to outline paths to act upon the constraints that force operators to carry out ineffective or detrimental forms of coordination and movement. However, acting on gestures is another thing altogether.

Gestures are, by nature, complex. They are contextual creations that allow operators to respond to the issues posed by the task. This involves taking very seriously the active role that the operator takes on in the production of a gestural solution. Therefore, in order to respond to task requirements, there is not just one, but several relevant gestural solutions, since the conditions and constraints that operators must cope with change over the course of their work: variability in products, in environmental conditions, in the state of operators, etc. One might add that the variability of gestures, for operators, constitutes a tool with three key functions: a function of effectiveness, a function of health preservation and a function of production of quality work. Acting on a gesture, therefore, implies liberating margins for manoeuvre, allowing operators to adjust their gesture and to repeat it without it ever being the same twice.

Liberating margins for manoeuvre, however, is a necessary but not a sufficient condition for the development of gesture. The development of a gesture requires opening a reflective space that allows operators to debate the gestural solutions that they choose to adopt. From this point of view, the variability, the obstacles encountered, just as the work collective, all play an important part in creating a reflective distance regarding one's gestures.

A constructive approach to ergonomics therefore implies acting at three levels: the level of constraints that make movements inefficient or detrimental, the level of margins for manoeuvre that make it possible to free up the space of gestural solutions adopted by operators, and finally, the level of the organization that allows debating and reflecting on the gestures used.

This perspective requires recognizing the intelligence and creativity of operators in their ability to invent new gestural solutions to respond to the demands of work.

References

Aptel, M., and Vezina, N. (2008). Quels modèles pour comprendre et prévenir les TMS? Pour une approche holistique et dynamique. Presented at 2ème Congrès Francophone sur les TMS, Montréal, Canada.

Barthe, B., and Quéinnec, Y. (2005). Work activity during night shifts in a hospital's neonatal department: how nurses reorganize health care to adapt to their alertness decrease. *Ergonomia IJE and HF*, 27(2), 119–129.

Bernstein, N.-A. (1967). *The co-ordination and regulation of movements*. Oxford: Pergamon Press.

Bril, B. (2012). Apprendre des gestes techniques. In E. Bourgeois and M. Durand (Eds.), *Apprendre au travail* (pp. 141–151). Paris: PUF.

Brisswalter, J., and Mottet, D. (1996). Energy cost and stride duration variability at preferred transition gait between walking and running. *Canadian Journal of Applied Physiology*, 21(6), 471–480.

Chassaing, K. (2010). Les "gestuelles" à l'épreuve de l'organisation du travail: du contexte de l'industrie automobile à celui du génie civil. *Le Travail Humain*, 73(2), 163–192.

Clot, Y., and Faïta, D. (2000). Genre et style en analyse du travail. Concepts et méthodes. *Travailler*, 4, 7–42.

Clot, Y., and Fernandez, G. (2005). Analyse psychologique du mouvement: apport à la prévention des TMS [version électronique]. *@ctivités*, 2(2), 68–78.

Coutarel, F., Daniellou, F., and Dugué, B. (2003). Interroger l'organisation du travail au regard des marges de manœuvre en conception et en fonctionnement. *Pistes*, 5(2), Retrieved from http://www.pistes.uqam.ca/v5n2/pdf/v5n2a2.pdf.

Davis, B., and Sumara, D. (2007). Complexity science and education: reconceptualizing the teacher's role in learning. *Interchange*, 37(1), 53–67.

Dugué, B., Chassaing, K., Coutarel, F., and Daniellou, F. (2010). L'ergonome peut-il contribuer à créer des systèmes adaptatifs et résilients? 5 ans après la conception d'une ligne de découpe, le retour sur un abattoir de canards gras. Presented at 45th Congress of SELF, Liège, Belgium.

Durand, M. (1993). Stratégie de recherche, optimisation et apprentissage moteur. In J.-P. Famose (Ed.), *Cognition et performance*. Paris: INSEP.

Fillietaz, L. (2012). Réflexivité et explicitation située de l'action des formateurs: une perspective interactionnelle et multimodale. In I. Vinatier (Ed.), *Réflexivité et développement professionnel* (pp. 275–304). Toulouse: Octares.

Kelso, J. A. S. (1995). *Dynamics patterns: the self-organization of brain and behavior*. Boston: MIT Press.

Leplat, J. (1995). À propos des compétences incorporées. *Éducation Permanente*, 123, 101–114.

Madeleine, P. (2010). On functional motor adaptations: from the quantification of motor strategies to the prevention of musculoskeletal disorders in the neck–shoulder region. *Acta Physiologica*, 199, 1–46.

Mollo, V., and Falzon, P. (2004). Auto- and allo-confrontation as tools for reflective activities. *Applied Ergonomics*, 35(6), 531–540.

Newell, K. M. (1986). Constraints to the development of coordination. In M. G. Wade and H. T. A. Whiting (Eds.), *Motor development in children: aspects of coordination and control* (pp. 341–360). Dordrecht, The Netherlands: Martinus Nijhoff.

Pastré, P. (1997). Didactique professionnelle et développement. *Psychologie Française*, 42(1), 89–100.

Pavageau, P., Nascimento, A., and Falzon, P. (2007). Les risques d'exclusion dans un contexte de transformation organisationnelle. *Pistes*, 9(2). Retrieved from www.pistes.uqam.ca/v9n2/articles/v9n2a6.htm.

Polanyi, M. (1969). *The tacit dimension*. New York: Doubleday and Company.

Rabardel, P., and Samurçay, R. (2001). From artifact to instrument-mediated learning. Presented at International symposium organized by the Center for Activity Theory and Developmental Work Research, University of Helsinki, Finland.

Rasmussen, J. (1983). Skills, rules and knowledge: signal, signs, symbols and other distinctions in human performance models. *IEEE Transactions and Systems, Man and Cybernetics*, 13, 257–266.

Rix, G., and Biache, M. J. (2004). Enregistrement en perspective subjective située et entretien en re-situ subjectif: une méthodologie de la constitution de l'expérience. *Intellectica*, 38, 363–396.

Schmidt, R. A. (1975). A schema theory of discrete motor skill learning. *Psychological Review*, 82(4), 225–260.

Schön, D. A. (1983). *The reflexive practitioner: how professionals think in action*. New York: Basic Books.

Sigaut, F. (2009). Techniques, technologies, apprentissage et plaisir au travail. *Technique and Culture*, 52–53, 40–49.

Simonet, P. (2011). *L'hypo-socialisation du mouvement: prévention durable des risques musculo-squelettiques chez des fossoyeurs municipaux.* Unpublished doctoral dissertation, CNAM, Paris.

Vermersch, P. (1999). Introspection as practice. *Journal of Consciousness Studies,* 6, 15–42.

Vézina, N., Prévost, J., Lajoie, A., and Beauchamp, Y. (1999). Élaboration d'une formation à l'affilage des couteaux: le travail d'un collectif, travailleurs et ergonomes. *Pistes,* 1(1). Retrieved from http://www.pistes.uqam.ca/v1n1/pdf/v1n1a3.pdf.

Vigarello, G. (1988). *Techniques d'hier et d'aujourd'hui.* Paris: Revue EPS and Michel Laffont.

chapter five

From constrained to constructed working time

Toward an enabling organization of work in rotating shifts and night shifts

**Cathy Toupin, Béatrice Barthe and
Sophie Prunier-Poulmaire**

Contents

In a context of constant expansion of atypical work schedules, in particular rotating shifts and night shifts, a systematic and constructive approach to ergonomics considers the temporal organization of work as a possible contribution to the quality of work, to the safety and reliability of systems, and to the health of men and women involved with these schedules.

First, we will describe the detrimental effects classically associated with the practice of rotating shifts and night shifts. This chapter will then aim to show that the organization of work schedules can also be respectful of the state of health of employees and can even, under specific conditions, contribute to the development of abilities, know-how and skills that are conducive to a successful professional career path. The joint analysis of the work carried out, of the individual characteristics of employees in charge of carrying out this work, of the strategies they constantly elaborate at work over months and years, makes it possible to imagine numerous paths for action. No doubt the detrimental effects of shift work will remain, but the organizational options chosen by companies may have a strong influence on these effects. Without cancelling them out, they can alleviate them by supporting the development of men and women at work, within a professional path that has been thought out and constructed beforehand.

This chapter will focus on rotating shift and night shift work. Indeed, these are, within the scope of all unusual work schedules, those that are the most widely used and which have been the strongest focus of attention, on scientific, economic, political and social levels. Rotating schedules are a direct consequence of shift work, which we will define as a mode of temporal organization of work in which several teams successively occupy the same workstations, at different times, to ensure continuity in goods or services.

Rotating shift and night shift work: A continuously expanding practice with detrimental effects

Although in France 37 per cent of all workers work following a 'normal' schedule – i.e. one that is close to the 'social day' (8 a.m.–6 p.m.) – nearly two out of three employees work schedules that can be said to be untypical, i.e. early in the morning, late at night, in rotating shifts, split shifts, long shifts (more than 10 hours), at the weekend, part-time, following unpredictable schedules, etc. (Bué and Coutrot, 2009). Thus, one of five employees works in a rotating shift or night shift team. In 2009, 15.2 per cent of employees (that is, 3.5 million people) worked at night, either habitually or occasionally; that is 1 million more people than in 1991. The proportion of people who stated they worked at night has more than doubled in 20 years (7.2 per cent in 2009 versus 3.5 per cent in 1991), with a stark increase for women (Algava, 2011). It should be noted that night-time work often adds up with other types of atypical schedules (rotating shifts, shifts varying from one week to the next, night work and work on Saturdays or Sundays).

At the level of the European Union, evolutions in past years have been quite uneven (in spite of a harmonization, at the European level, of labour legislation concerning night work), with night work declining in countries neighbouring France in recent years. However, according to an inquiry carried out by the European Foundation of Dublin, the percentage of employees engaging in night-time work varies between 18 and 24 per cent in the 31 participating countries (Edouard, 2010).

These kinds of temporal organizations of work place individuals in situations with conflicts of temporalities (Barthe, 2009; Quéinnec et al., 2008). This may have detrimental effects on work, but also on the health and the family and social lives of the persons involved.

Indeed, rotating schedules – and particularly night schedules – can lead to adverse consequences on work, notably in terms of safety and reliability. The times at which have occurred the most serious industrial accidents and catastrophes of the past century lead us to question this relationship: Three Mile Island (1979) at 4 a.m., Chernobyl (1986) at 1:30 a.m., Bhopal (1984) at 12:45 a.m. and the gas explosion at a Total refinery (1992) at 5:22 a.m. At a different scale, a scientific study of 1020 fatal occupational accidents in Australia showed that the mortality rate is two times greater in night shifts than in day shifts (Williamson and Feyer, 1995). Folkard and Tucker (2003), based on several studies of industries operating in three-shift systems, have shown that compared with the morning shift, the afternoon shift is associated with an 18.6 per cent increase in the probability of an occupational accident, and the night shift with a 30.4 per cent increase. Folkard (1981) summarized the results of a set of studies showing variations in performance in the course of the 24-hour day, with a marked decrease between midnight and 4 a.m. in various professions (e.g. longer response times in telex operators, rise in errors of meter reading in factories, drowsiness while driving, lack of responses of train drivers to traffic signals, increase in the rate of accidents in hospitals, etc.). The variation in observed performance is close to the variations in the wakefulness of workers, following a circadian rhythm over the 24-hour clock, with a minimum in the middle of the night and a maximum during the afternoon.

In addition to the effects of these temporal organizations on reliability and safety of and at work, impacts on health are numerous and undeniable: a quantitative and qualitative deterioration of sleep, more or less severe disruptions of digestive functions, nervous disorders potentially leading to a depressive state, rise in the risk of cardiovascular disease, etc. (for a summary, see Gadbois, 1998), pregnancy disorders (Croteau, 2007) and cancer risks, notably of the breast and colon (Haus and Smolensky, 2006). Hence, in France, work in rotating schedules and night schedules has been considered detrimental since the Retirement Reform of 2010.

One should note that the prevalence of health disorders in salaried workers is variable and depends on the exposure time, on what scheduling systems are being practiced, on the characteristics of the work situations, on constraints related to personal and family life, age, etc. Some people tolerate quite well the practice of rotating shifts and night shift work, whereas others must quickly abandon them. Yet, returning to a regular daytime schedule does not necessarily imply that those health disorders will disappear (Bourget-Devouassoux and Volkoff, 1991).

Employees working in rotating and night schedules, finally, find it difficult to cope with the discrepancies between professional life and social and family life, because of the conflicts existing between these schedules and the need to have some time available to share activities with one's family and friends. Consequences may occur at several levels: relationship difficulties, a decrease in the frequency and quality of time spent with one's children, a decrease in the time spent working in an association and with friends, 'social isolation', etc. (Prunier-Poulmaire, 1997).

From constrained time to constructed time

Hence, practices in work scheduling may lead to hazards to employees and, in particular, contribute to a decline in their state of health – but also contribute to disruptions in their personal life.

However, although rotating schedules and night schedules require the employees involved to work against their normal physiological, psychological and social modes of operation, we also know today that they do not remain passive when confronted with difficulties inherent to this kind of organization. Employees implement processes of adaptation at work, which are constructed and refined over months and years of practice.

Thus, in specific conditions, the design of work schedules can contribute to 'constructing the state of health' of operators – or at least to not harming it (Gollac and Volkoff, 2007). The organization of work schedules can also be a source of self-development, learning, acquisition of skills and autonomy, knowledge, know-how, development of regulatory strategies of possibilities of learning from oneself and from others, and therefore of developing health in a broad sense.

Constructed time: Strategies at work to preserve health

A series of ergonomic studies has cast light on the ways in which operators, when subjected to the circadian variations of their own psychophysiological functions, cope with a decrease in vigilance in order to achieve their work goals. These studies show the construction and implementation of adaptations that manifest themselves in the work activity itself, in a quantitative and qualitative restructuring of activity, that can be perceived at individual and collective levels (Barthe et al., 2004). Operators work differently during the day and at night, as well as over the progression of a night shift, without this affecting their effectiveness at work. These regulations have been observed in numerous professional sectors: process control, satellite control, chemical and petrol industries, daily press, the hospital sector, the transport sector, etc.

Part of the variations observed in work activity directly reflect the level of functional activation of individuals. This quantitative variability, observed in some dimensions of activity throughout night shifts, has notably been noted with regard to communication at work, movement and information gathering. These findings demonstrate a gradual decrease in activity throughout night-time work, reproducing a curve similar to the circadian variation of wakefulness with a minimum between 1 a.m. and 3 a.m.

However, performing more actions or accomplishing them faster during a decrease in wakefulness does not imply working less effectively. Adjustments are made so that the crucial goals of activity are reached at all times, and so that productivity is identical. In addition to the quantitative variations mentioned above, one must also add qualitative reorganizations of work activity. For example, surveillance personnel working a night schedule will group together the tasks that require reflection, precision and decision-making in the earliest part of the night, and then introduce the tasks that are more physical and do not require quite so much attention (Prunier-Poulmaire, 2008). This strategy makes it possible to maintain wakefulness and to break up the monotony of some of the tasks. In the hospital sector, nurses and pediatric assistants, over the course of long night shifts (11.5 hours), use strategies for care that are both specific and quicker at 2 a.m. compared with other periods of night-time care. Some secondary care activities are postponed in order to preserve the sleep of infants, as well as avoid the accumulation of fatigue of the caring staff, by allowing them to take a break before the final stage of the care schedule, which they view as the most difficult (Barthe and Quéinnec, 2005).

All of these results focus here on the activity of an operator alone. However, collective reorganizations can also occur within teams, in order to cope with the requirements of work while collectively managing the individual variations of the level of wakefulness. Dorel and Quéinnec (1980) showed that in the control of a production process in a drinking water factory, there was a collective reorganization of the supervision work aiming to give as much responsibility as possible to the operator who will have a 3-day rest period at the end of the night. This makes it possible to protect the second operator, who will return the following night. In the neonatology unit mentioned above, the nurses and assistants assist each other in order to decrease their workload and provide each other with specific technical help or skills at various times during their schedule (Barthe, 2000). They also organize their work collectively so as to be able to allow themselves individual breaks. In customs brigades, an informal attribution of tasks between agents is implemented. It is the agents who begin the night shift who take on the most delicate and hazardous tasks (e.g. interrogating users who have committed an offence), since these require a great amount of self-control, attentive listening, boundless patience and a strong concentration (Prunier-Poulmaire, 1997).

It seems therefore legitimate to claim that one can 'learn' to work by night or in rotating schedules, or at least learn to 'skilfully deal with it'. One can acquire specific knowledge about oneself and one's tasks – modulating these tasks differently depending on at what time they are carried out. This knowledge makes it possible to reorganize work. The strategies presented above, which in concrete terms are translated notably into a temporal reorganization of specific actions occurring at night, suggest that workers have strong needs in terms of autonomy.

A construction over the course of months and years

Over the course of months and years spent working atypical hours, workers develop specific experience, skills, knowledge and know-how for the work carried out in rotating schedules and night schedules. This experience provides workers with resources allowing them to better manage the difficulties and requirements that are specific to their schedule, or to protect themselves from them.

A study conducted in the hospital sector (Toupin and Volkoff, 2007; Toupin, 2012), involving night shift nurses in a pneumology department, illustrates this point. The observations and analyses carried out as part of the ergonomic intervention aimed to highlight the ways in which experience allows the nursing staff to better 'manage the night'. In this sector, the work that is carried out is not very different between night and day, but it does have some specific features in the night-time. The conditions in which the prescribed tasks are achieved are particular, because of the psychophysiological state of the nurses (fatigue, decrease in wakefulness), of the features of the work environment at that time (work in small groups, doctors and middle managers are absent, etc.) and of the state of patients (tired, anxious, etc.). This specific character of nocturnal activity must be highlighted, in order to avoid considering the night nurse – whose work is tightly prescribed, with compulsory tasks that are entirely dictated by medical prescriptions and by the state of patients – just like a nurse who would simply work in another period of the nycthemeral clock. This is also true in many other professional sectors (Prunier-Poulmaire and Gadbois, 2004).

Yet, workers who start working the night shift are not always sufficiently informed and trained with respect to the specific features, stakes and difficulties of the trade in this period of the day. This may lead to problematic situations. For example, how can one manage a decrease in vigilance during the night shift and the appearance of fatigue during a cycle of rotations over several successive work nights, during emergency situations that require the worker to be alert and awake?

Over years of practice, night workers redefine their tasks, by setting themselves new, 'temporally situated' goals (Gaudart and Ledoux, this volume). With experience comes the wish to anticipate the future work activity, with two main goals:

- To limit and cope with the fatigue that emerges during the night. For example, in the hospital sector, nurses seek to avoid carrying out some (physically and cognitively) demanding, delicate or hazardous tasks, at a time where they know, out of experience, that they are less vigilant. When this is not a detriment to the health of the patient, they can elect to put forward or postpone an action so as to allow themselves a resting period in the middle of the night. Forms of cooperation also appear in the team when there is a need to move or change a patient, or whenever a nurse is feeling too tired or not alert enough to take good care of a patient: for example, when there is a need to jab a patient infected with HIV (Toupin, 2012) or to replace a drip on a premature baby (Barthe, 2000).

  In a very different sector, customs agents elect to carry out, in the beginning of the night, controls in strategic locations that they know, from experience, to be most likely to present a hazard (Prunier-Poulmaire et al., 1998). Conversely, they keep the areas that require a lower level of vigilance for the later part of the night.

- To limit and avoid emergency situations, sources of fatigue and stress (notably because of the lack of management) in order to have, whenever possible, a work activity that is better managed. In a steel mill, part of the actions aiming to assess the quality of reels are carried out beforehand. Controlling one reel will lead workers to decide which controls they will perform two or three reels down the line. These control anticipations are more frequent when the workers are experienced, and when the work is carried out at night. This mode of operation allows workers to avoid working in an emergency, at a time where their wakefulness may be reduced and their memory less effective (Pueyo et al., 2011).

  In the hospital, nurses make sure, at the beginning of the night shift, that the medical prescriptions will allow them to deal with potential anxiety attacks of the patients who they will be responsible for during the night (Toupin and Volkoff, 2007; Toupin, 2012). The level of anxiety of the patient is a very important parameter, since it affects how the night will play out – for the patient and for the team of caretakers, particularly the nurse in charge of that patient, but also for other patients in the department who, following the call of an anxious patient, may wake up, possibly feeling in pain or anxious themselves. Nurses also make a point that the first round should be done early, so that they can see as many patients awake as possible, construct a representation of their state of health, and imagine how the night is going to play out for the patients and for themselves.

Similarly, in process control, controllers collect twice as much information per unit of time at the start of their shift than during the rest of their shift (Andorre and Quéinnec, 1996). This intense and global process of information gathering allows them to update their representation of the state of the system, and to have access to knowledge in order to be able to anticipate future variations.

As a general rule, various sources of experience can support these evolutions, leading workers to alter their practice over the course of months and years:

- Experience derived from professional practice: Once they have been confronted on several occasions with problematic situations during the night shift, without the possibility of calling upon management, operators, for example, alter the ways in which they ask questions during oral feedback with the afternoon team, or reallocate differently the tasks that they have to complete during the night.
- Experience derived from the practice of colleagues, viewed as a source of communication and learning: Operators who are beginning night shift work often claim that they draw inspiration from the modes of operation and strategies implemented by their more experienced colleagues, from the advice they are given, from the know-how that is shared, and by the abilities that are developed within and for work activity.
- Experience derived from the knowledge of oneself, from psychophysiological abilities, from the assessment of one's own state of health during the work, and the impact of this state of health on the ability to stay awake and react promptly and effectively during the entire shift.

Thus, the conditions of nocturnal activities do not only cause the same task (as is carried out during the day or following regular hours) to be performed differently. It is *another* task that is being performed, because of the specific requirements related to the night shift (e.g. increased liability and autonomy) and of a different weight in the importance of some criteria (e.g. variations in task requirements, management of fatigue, etc.). These elements lead to operators rethinking their work in different terms, and constructing new skills that will have an influence – on the arduousness of work and on the preservation of health, on the one hand, and on the quality of work, on the other hand.

Designing enabling organizations for work in rotating schedules and night schedules

When one takes an interest in the organization of work schedules, the possible means for action undoubtedly rely on the design of schedules that are compatible with current medical knowledge (Folkard, 1992; Knauth, 1996) – but not only that. One should also rely on the focus of the sections above: implementing work organizations that are mindful of operators' individual and collective strategies, which allow and indeed support the development of their practices, experience and skills. Such development is a contribution to the quality of work, but also to the preservation of workers. It should be encouraged and be a goal of ergonomic action. From there, several levels of action can be envisioned.

Acting on the conditions and contents of work during these schedules

Quite often, work in rotating schedules and night schedules combines with other constraints: temporal, physical, environmental, psychical, organizational, etc. (Algava, 2011; Bué and Coutrot, 2009; Volkoff, 2005). Alleviating these constraints then makes it possible to improve the conditions of work, and therefore the health of the people working in these organizational modes of scheduling. First, it is crucial to reflect on the nature of the tasks attributed to operators, in order to reduce the weight of the constraints that are contained within rotating shift and night shift work. In particular, the goal is to design tasks that are compatible with the functional abilities – both physical and cognitive – of the workers, taking care to reduce constraints related to rhythm, physical effort, attention, memorization, etc. One could also imagine reorganizing specific night-time tasks – or even transferring these tasks to the daytime – and to offer judiciously spaced breaks over the course of the shift.

The knowledge we have about real-world work, about the individual and collective adaptation processes implemented at certain times in the shift, allows us to imagine some paths for future reflection.

The design of work conditions aims to introduce more flexibility in the prescribed task, in leaving employees with operational leeway and autonomy (based on the individual and collective possibilities for regulation) that is both sufficient and acceptable from the point of view of health and safety, so that they can organize their activity. This autonomy constitutes a true goal for the design of work schedules. The goal is not to tolerate autonomy but to support and construct it.

As we have seen, in the context of work in atypical and night-time schedules, the collective is an important resource (cf. Caroly and Barcellini, this volume). It therefore seems crucial to reflect on the number of employees in work teams and on the characteristics of the workers, from the point of view of their skills and know-how, notably in order to support exchanges and learning between colleagues. A sufficient number of colleagues would also make it possible to implement official break times during shifts, notably over the course of the night by ensuring the sharing of skills within the team. Night-time naps are not a common practice in most countries. Yet, it has beneficial effects on the level of vigilance and fatigue at the end of a shift and during a shift rotation cycle (Matsumoto and Harada, 1994), as well as on mood (Kaida et al., 2007) and on some of the operators' cognitive functions, in particular on the risk of making errors (Bonnefond et al., 2001).

It is also necessary to ensure that the periods of co-presence between arriving teams and departing teams are long enough to facilitate the sharing of information and the gradual takeover by arriving operators (Le Bris et al., 2012).

Finally, supporting the construction of experience (practices and skills), for example, through opportunities to learn, train and reflect together about the nature of the work, is a primary stake of ergonomic interventions. Exposing the activity carried out during rotating schedules or night schedules may serve as a starting point to the recognition of the specificities of various professions during those times, and to the consolidation of learning. Furthermore, reflective activities (Mollo and Nascimento, this volume) may contribute to constructing and developing the nocturnal skills of employees and their capacity for action. To achieve this, however, the work system (the composition and stability of collectives, the opportunities to think together about work, the training programs, etc.) should support the construction of this experience and the possibility of using it at work.

Acting on professional careers and human resources management

As we have seen, some characteristics of work situations can support or hinder the construction of experience. However, the question is not to focus exclusively, in a context where professional careers are increasingly chaotic, on the characteristics of work situations at a given moment, but to imagine their succession over the years. Time, then, must be understood not at the level of the company, but at the level of a career; here it is crucial to adopt a diachronic view.

It is all the more important to study work in rotating and night shifts, in its relationship with career paths, ageing and the construction of experience, because the current sociodemographic context is marked by a dual trend: the ageing of the active population and the increasing prevalence of atypical and night shifts. These trends should have two consequences: an increase in the prevalence of atypical shifts (notably night shifts) in ageing employees – which one can already observe today in France – and increasingly frequent situations of employees from different generations working alongside each other. In this context, issues of health and experience deserve a close scrutiny, and preventive actions should be carried out.

Thus, means of action are not located solely at the level of work situations, but also at the level of career paths (Prunier-Poulmaire et al., 2011). These rely in particular on reducing the duration of exposure to work in rotating and night schedules, and to atypical schedules in general (shorter schedules, part-time work, introduction of break periods in the workplace or at home, reduction of the time spent in the work situation, etc.).

Designing the working time therefore involves anticipating it better, by implementing better means of career management, regular and frequent medical checkups on employees working shift work and night work, ensuring access to on-the-job training, etc. By doing this, the goal is to stop the potential exclusion of some employees from their workstation, from their company or their establishment, or even from the job market.

Conclusion: From constrained time to constructed time

Regarding the issues of time and work, the ergonomic approach is undeniably an asset. It is advisable to focus not only on work schedules as such, but also on the persons who are subjected to those schedules, to the characteristics of the work that is entrusted to them, and never to consider these elements independently from one another, following a strong principle of inseparability (Prunier-Poulmaire, 1997). Designing the work without thinking about what is at play within this work, what occurs in this work, what is done in this work, makes no sense.

Therefore, and surprisingly, designing work schedules involves first carrying out fine-grained analyses of the work activity – the very activity that is to be carried out within the confines of the proposed time. These analyses make visible what was invisible, that is, the major differences in strategy between day and night, that allow employees to confront the specific requirements of their work. It is in this way that ergonomic analysis takes on its full meaning; by analyzing the activity being considered, by taking into account the singularity of the work situation that is studied and the specific features of the populations considered, we can reveal the individual and collective strategies whose importance we have highlighted over the course of this chapter.

Furthermore, designing work schedules, particularly when they are rotating or night shifts, involves ensuring they will allow employees to learn about themselves, about other people in the collective, about the work itself, and about its variability in the hours of day and night. To be able to develop original know-how and specific skills requires being able to understand these things. These acquisitions, these assets, can all be transferred from here to elsewhere, constituting resources in the construction of a chosen career path. Indeed, if one accepts that professional experience is not limited to capitalizing mastery of specific tasks, but also includes a knowledge of the contexts in which these resources are deployed (Pignault and Loarer, 2008), then the knowledge of these specific temporal contexts, of the particular abilities one develops within them, are all resources that could be transferred to support career mobility in a way that is constructed and chosen by the worker. At this point, the practice of rotating work schedules and night schedules could be seen as a resource for the future.

Types of organization should also support margins for manoeuvre at work – for example, by allowing the tasks to be carried out to be decomposed or spread out over time – and make it possible to choose the more efficient modes of operation, i.e. those that are most compatible with the internal state of the worker. Hence, setting up the organizational conditions that are favourable to the processes whereby these strategies are constructed, providing genuine latitude for decision and for action, equates with contributing to the development of skills in operators and to the preservation of their health. Finally, designing work schedules implies providing a temporal context that allows and encourages autonomy – the very autonomy that actively contributes to the development and the construction of health.

As we can see, the definition of a *favourable* schedule is a complex matter. Yet, it remains an important issue because it contributes to setting up an 'enabling environment' – that is, an environment that is mindful of individuals in the here and now, but also that is likely to preserve their abilities in the long run, abilities for future action. Ergonomics can then contribute to designing the time spent at work, so that it is not a *constrained time*, but a *constructed time*.

References

Algava, E. (2011). Le travail de nuit des salariés en 2009. Fréquent dans les services publics; en augmentation dans l'industrie et pour les femmes. *DARES Analyses*, 9.

Andorre, V., and Quéinnec, Y. (1996). La prise de poste en salle de contrôle de processus continu: approche chronopsychologique. *Le Travail Humain*, 59, 335–354.

Barthe, B. (2000). Travailler la nuit au sein d'un collectif: quels bénéfices? In T. H. Benchekroun and A. Weill-Fassina (Eds.), *Le travail collectif. Perspectives actuelles en ergonomie* (pp. 235–255). Toulouse: Octarès.

Barthe, B. (2009). Les 2x12h: une solution au conflit de temporalités du travail posté? *Temporalités*, 10. Retrieved from http://temporalites.revues.org/index1137.html.

Barthe, B., and Quéinnec, Y. (2005). Work activity during night shifts in a hospital's neonatal department: how nurses reorganize health care to adapt to their alertness decrease. *Ergonomia IJE and HF*, 27(2), 119–129.

Barthe, B., Quéinnec, Y., and Verdier, F. (2004). L'analyse de l'activité de travail en postes de nuit: bilan de 25 ans de recherches et perspectives. *Le Travail Humain*, 67(1), 41–61.

Bonnefond, A., Muzet, A., Winter-Dill, A. S., Bailloeuil, C., Bitouze, F., and Bonneau, A. (2001). Technical note – innovative working schedule: introducing one short nap during the night shift. *Ergonomics*, 44(10), 937–945.

Bourget-Devouassoux, J., and Volkoff, S. (1991). Bilans de santé et carrière d'ouvriers. *Economie et Statistique*, 242, 83–93.

Bué, J., and Coutrot, T. (2009). Horaires atypiques et contraintes dans le travail: une typologie en six catégories. *DARES Premières Synthèses*, 22(2).

Croteau, A. (2007). *L'horaire de travail et ses effets sur le résultat de la grossesse. Méta-analyse et méta-régression.* Institut National de Santé Publique du Québec. Retrieved from http://www.inspq.qc.ca.

Dorel, M., and Quéinnec, Y. (1980). Régulation individuelle et interindividuelle en situation d'horaires alternants. *Bulletin de Psychologie*, 33(344), 465–471.

Edouard, F. (2010). *Le travail de nuit: impact sur les conditions de travail et de vie des salariés.* Report from the Economic, Social and Environmental Council. Paris: Conseil Economique, Social et Environnemental.

Folkard, S. (1981). Shiftwork and performance. In L. C. Johnson, D. I. Tepas, W. P. Colquoun, and M. J. Colligan (Eds.), *Biological rhythms, sleep and shiftwork, advances in sleep research* (Vol. 7, pp. 283–305). Lancaster, UK: M.T.P. Press Ltd.

Folkard, S. (1992). Is there a "best compromise" shift system? *Ergonomics*, 35(12), 1453–1463.

Folkard, S., and Tucker, P. (2003). Shift work, safety and productivity. *Occupational Medicine*, 53, 95–101.

Gadbois, C., (1998). Horaires postés et santé. In M. Lenoble (Ed.), *Encyclopédie medico-chirurgicale: toxicologie-pathologie professionnelle* (16-785-A-10). Paris: Elsevier.

Gollac, M., and Volkoff, S. (2007). *Les conditions de travail aujourd'hui.* Paris: La Découverte.

Haus, E., and Smolensky, M. (2006). Biological clocks and shift work: circadian dysregulation and potential long-term effects. *Cancer Causes Control*, 17, 489–500.

Kaida, K., Takahashi, M., and Otsuka, Y. (2007). A short nap and natural bright light exposure improve positive mood status. *Industrial Health*, 45, 301–308.

Knauth, P. (1996). Designing better shift systems. *Applied Ergonomics*, 27(1), 39–44.

Le Bris, V., Barthe, B., Marquié, J. C., Kerguelen, A., Aubert, S., and Bernadou, B. (2012). Advantages of shift changeovers with meetings: ergonomic analysis of shift supervisors. *Applied Ergonomics*, 43(2), 447–454.

Matsumoto, K., and Harada, M. (1994). The effect of night-time naps on recovery from fatigue following night work. *Ergonomics*, 37(5), 899–907.

Pignault, A., and Loarer, E. (2008). Analyser l'expérience en vue de la mobilité professionnelle: une nouvelle approche. *Éducation Permanente*, 174, 39–50.

Prunier-Poulmaire, S. (1997). Contraintes des horaires et exigences des tâches: la double détermination des effets du travail posté. Santé et vie socio-familiale des agents des Douanes. Unpublished doctoral dissertation, Ecole Pratique des Hautes Etudes, Paris.

Prunier-Poulmaire, S. (2008). Horaires décalés. Salariés à contretemps. Concilier horaires et activité. *Santé et Travail*, 61, 30–31.

Prunier-Poulmaire, S., and Gadbois, C. (2004). Temps et rythme de travail. In E. Brangier, A. Lancry, and C. Louche (Eds.), *Les dimensions humaines du travail.* Nancy, France: Presse Universitaire de Nancy.

Prunier-Poulmaire, S., Gadbois, C., Ghéquière, A., and De La Garza, C. (2011). Volver a pensar la organización del tiempo de trabajo cuando la tecnología cambia: el caso del equipo de operación de una central nuclear. *Laboreal*, 7(2), 10–24.

Prunier-Poulmaire, S., Gadbois, C., and Volkoff, S. (1998). Combined effects of shift systems and work requirements on customs officers. *Scandinavian Journal of Work, Environment and Health*, 24(3), 134–140.

Pueyo, V., Toupin, C., and Volkoff, S. (2011). The role of experience in night work: lessons from two ergonomics studies. *Applied Ergonomics*, 42(2), 251–255.

Quéinnec, Y., Teiger, C., and de Terssac, G. (2008). *Repères pour négocier le travail posté*. Toulouse: Octarès.

Toupin, C. (2012). L'expérience du travail de nuit chez des infirmières de pneumologie. In C. Gaudart, A. F. Molinié, and V. Pueyo (Eds.), *La vie professionnelle – age, expérience et santé à l'épreuve des conditions de travail* (pp. 161–177). Toulouse: Octarès.

Toupin, C., and Volkoff, S. (2007). Experience and compromise in night shifts among hospital nurses. *Ergonomia*, 29, 137–142.

Volkoff, S. (2005). *L'ergonomie et les chiffres de la santé au travail: ressources, tensions et pièges*. Toulouse: Octarès.

Williamson, A., and Feyer, A. (1995). Causes of accidents and the time of day. *Work and Stress*, 9(2–3), 158–164.

chapter six

Activity as a resource for the development of work organizations

Fabrice Bourgeois and François Hubault

Contents

Any mobilization of resources in order to act in a given context requires the implementation of a work organization. Ever since Taylorism and Fordism, the form favoured by work organizations has been the proceduralization of work, including in the case of newer, more recent forms of work organizations such as Lean manufacturing.

Such an approach posits that the effectiveness of the organization resides in the value of the procedures that it implements to act in the real world, and in the control of the organization over how these procedures are applied. Yet, ergonomic analysis shows that in order to be effective, the activity of workers – this term also includes managers – drifts away from strict adherence to these rules. This does not, however, condemn the principle of organizing work, but differs from a deterministic approach to organizations, which confuses anticipation with reality. This approach forces workers to cope with insufficient or ill-suited means, at the risk of disrupting their effectiveness and their health.

In this chapter, we will defend a positive view of activity as a resource for the organization. This vision implies redefining the role of standardization and the nature of prescription, and specifying the nature of the relationship between the organization and the real world, on the one hand, and between the organization and subjectivity, on the other hand.

When standardization freezes the organization, the workers are put to trouble

Taylorism has claimed to 'scientifically' prescribe the best way for all to perform a job. From the start, the Taylorian organization denies any legitimacy of the worker in deciding the worth of his or her own empirical way of working, entrusting company management with supervising the work, and the engineering sciences with designing and prescribing work. This repossession is one of the keys to the industrial revolution. This model of effectiveness is said to be scientific because it relies on a formal correspondence between measures of physical activity and industrial performance. Through Fordism, it allowed the development of employment and purchasing power for many people who did not have access to them or who wished to leave the social condition associated with agricultural work. However, criticism of this approach emerged quite quickly. American workers' unions blamed it early on for turning humans into machines, as evidenced by Taylor's famous phrase to workman Shartle: 'You are not supposed to think. There are other people paid for thinking around here'. Closer to our own history, many industrial actions in the 1970s denounced the effects of Taylorism: monotony, repetitiveness and a lack of prospects.

Activity ergonomics was born in this context, based on three main critiques of Taylorism and Fordism:

- On the one hand, they reduce the mobilization of the 'human resource' to the *execution* of simplified and predefined gestures (division of labour).
- On the other hand, neither understands that *the real world as it imposes itself to the workers does not fit in the division of labour as it is initially planned,* in such a way that workers must cope with situations where resources are sparse.
- Finally, they both oppose *drifts from procedure* by likening them to defects (errors, violations), denying that they might be a source of any effectiveness and blaming them for detrimental effects on physical and psychical health.

Therefore, activity ergonomics is opposed to a certain way of organizing work when it observes that the strict application of standards is ineffective.

This was pointed out even in the first ergonomic interventions ever performed in France, such as the intervention in 1969 in a television manufacturing company (Laville et al., 1972). Workers picked small television parts from about 30 boxes placed in front of them (e.g. wires, resistors, diodes, capacitors) of all shapes, colours and sizes before inserting them within a 90-second time span into small holes on a metal 'plate', advancing at a speed of 1 meter per minute on a conveyor belt. At the time, the dominant critiques of Taylorism – particularly that developed as early as 1956 by Georges Friedmann – pointed out the monotony of work related to the repetitiveness of gestures. Indeed, this criticism denounced the effects, but not the supposed equivalence between the work that is supposed to be done (the standard) and the work that is done in reality (the activity), which it therefore seemed to validate. Yet, researchers in ergonomics discovered something else. The modes of operation implemented by the female workers, in fact, coped with a wide variety of situations where the small parts were inserted and extracted. These situations were related to quality defects or difficulties in separating the components. All of them mobilized a more sustained attention, complex cognitive processes and more intense muscular effort than had been 'planned'. The rhythm had been calculated for supposedly simple gestures, and not to manage these difficulties. The impact of the mobilization on the health of workers had become much greater as a result (Teiger, 2008).

These observations and explanations are still valid today, and they still upset the convictions of most economic decision-makers. According to them, the standardization of prescribed work justifies the relevance and effectiveness of investments. To ergonomics, however, standardization as a project is not relevant, since it constitutes a limitation or even a hindrance to workers seeking to achieve expected results (process safety, quality, time, etc.). Obtaining the expected results may rely on automation, if the production system is stable enough to be entirely formalized; failing that, it must rely on a system where humans deal with the part of the system that is not or cannot be mastered, using informal rules that they are in charge not of applying, but of mobilizing in order to manage the risks (de Terssac, 2012).

Such observations are all the more topical in a service-based, immaterial economy. In the dynamics of a service*, the 'beneficiary' – i.e. a customer,

---

* It is important to make a distinction between *service sectors* and the *service approach*, which can apply to any activity – agricultural, manufacturing, services; in other words, to distinguish the industrial logic (which can permeate down to the service sector) from service-based logic (which may inspire evolutions in the organization of the industry itself) (du Tertre and Hubault, 2008).

a patient, a user – cooperates in the production of a service that he does not merely purchase and consume, but also co-produces or even co-designs. From there, the sources of prescription become increasingly complex. They are no longer just vertical and hierarchical (i.e. work that is prescribed by management), but they are also lateral and transverse (i.e. the expectations of customers, colleagues, etc.). The concrete production of value thus depends on the ability of workers to carry out real-time arbitration between multiple prescriptions. Let us take the example of a gathering of workers in front of car bumpers leaving a painting line in a factory that produces automotive equipment. According to standard procedure, they should be standing in a line, concerned with detecting grains and sanding them down before the bumpers are sent to the end customer. Why are they exchanging opinions? This morning, the amount of grains is very high and occurs too often over the bumpers. They will obviously plan a longer sanding period, mostly to sand down the visible traces, as well as customer returns because of quality dropouts. But they hesitate to send the bumpers back to the start of the painting line because of the delays this will cause to production and delivery. Because of this, after some brief exchanges they decide to call the quality manager, who comes in and decides to accept the defects to satisfy the criterion of volume within the time set in the contract. Based on a strategic appreciation of the situation, where his knowledge of the immediate needs of the customer plays a crucial part, he would probably have decided differently if customer needs had seemed different to him. Indeed, this is what the concept of customer implies: understanding needs beyond what has been contractually and formally planned in such a way that the service relationship goes necessarily beyond what the agreement had planned. The workers were conscious of the need to assess the relevance of prescribed work and to allow themselves some leeway concerning the criteria, but lacked the authority to arbitrate between each other. The deliberation, when faced with the product, suggests an ability to clinically analyze the situation. The goal is related both to performance in activity (i.e. doing a good job) and to their health (having something to do with this useful work), but also to an impossibility of action. This is the very essence of the concept of service: knowing and being able to decide the level of quality, and thus economic performance, that is acceptable. From this point of view, one should note that the 'modernization' of public services most often follows – or, one should say, imitates – the dominant industrialist solution that is implemented in large private companies. This tends to produce the opposite effects to those that had been claimed by the advocates of new public management: declines in the quality of service, psychosocial tensions in the activity of 'agents' and rudeness on the part of 'beneficiaries'.

Work activity as a resource for organizing work and of organized work

Work activity takes charge of the drift between what is planned and what really happens. Because of this, it serves many purposes by adapting the organization of work when and where this organization is not able to deal with the real world as it is. Thus, the activity is both *organized* by the prescriptions and a *reorganizer* of the system as it was initially planned (cf. Arnoud and Falzon, this volume).

Activity as a resource for organizing work

In order for the work organization to take into account the real-world activity, it must deal with an apparent contradiction. Indeed, when designing the organization, activity, by its very nature, cannot be known beforehand; nor can it be prewritten as a standard might be. Its integration within the work organization can only be imagined in potential terms. Activity represents what the worker will have to/be able to/know how to put into play to carry out what is expected. Hence, it is the process – more than the 'done deal' – that the organization can deal with. This echoes the concept of 'work of organizing' proposed by de Terssac.

From there, the value of activity is in its *relevance*. Activity only reveals itself in real life, during an encounter with the real world where it will reveal its potential. By definition, the real world is not completely known in advance either. At best, its contents are supposed (i.e. the organized work in Figure 6.1). During its future encounters with the real world, work activity will take charge of the discrepancies between this idea of the future and the real-world situations. This is what we call the work of reorganization. The whole issue, when designing a work organization, is to

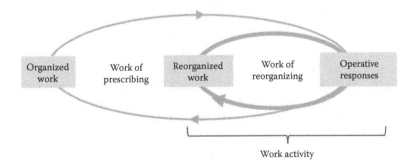

Figure 6.1 From the organization of work to the work of organizing.

produce elements of knowledge regarding this future activity, so that the prescribed work may include some leeway, allowing the implementation of several operating strategies to cope with the probable variety of real production situations. Anticipating the probable diversity of these situations in the future, as well as the breadth of modes of operation that are necessary for dealing with these situations, relies on methods aiming to simulate probable future activity (Daniellou, 2004) and probable future organizations (Van Belleghem, 2012; Barcellini et al., this volume). While seeking not to standardize everything and to highlight the need for plasticity in any work system (Béguin, 2007; Hubault, 2004), these approaches invite us to view work activity as a resource to continue – i.e. to realize, in real-world situations of use, the design of systems implemented by the organization, which is, by definition, unfinished (Vicente, 1999).

Organizational prescription and organized work both emanate from the 'cold world'. This is the world of the projections that are made in the design stage, based on the scenarios that are imagined relative to the future and on the experiences remembered from the past. Activity, on the other hand, develops in the emerging world where it produces other rules, based on every real-world situation. However, to achieve this, the work organization must allow autonomy in individual workers and in the collective of workers, within a framework that is plastic enough to allow deformation and reformation (Maggi, 1996). The confrontation between the initial rules and the rules produced by individual workers and collectives of workers requires some adjustments. The evaluation of these adjustments by individuals and collectives depends on the performance criteria that they give themselves. This 'hot' evaluation of the issues raised by a strict adherence to the rules occurs in live action. It must then 'cool down' through a process of collective deliberation. This can occur either between peers or with the hierarchy – i.e. through horizontal or vertical collaboration (Dejours, 2009). This allows the adaptations implemented by the workers in real time to cease to be an object of illegitimacy and conflict (de Terssac, 2012). This is exactly the goal of resource management: to be able to accept the immaterial investments that are necessary to capitalize on this know-how (issues of quality, safety, innovation, development, etc.). It is these very issues that form the main part of competitiveness in modern economies (du Tertre, 2007). Regarding this, let us note that ergonomic analyses tend to converge with economic, sociological or psychological analyses – all of which have a similar relationship to prescription.

Activity as a resource of organized work

For activity ergonomics, work organization is a dynamic system where the initial prescriptions, real-time reorganizations and cooled down adjustments based on earlier activity constantly combine with one another.

This view of organizations and prescriptions:

- Provides a response to the variability of situations. It would be point-less to attempt to make this variability disappear entirely.
- Proposes an approach of standards as a reference base that workers can refer to, to access operative possibilities that have been exper-imented on in the past. In this respect, it suggests rethinking the usual relationship to standards, making it possible to act in an open world, which is both varied and subject to variation.
- Produces an operative framework of rules that workers can find their way in. This framework supports the sense of belonging to a collective and to a trade.
- Supports making connections between the goals workers set for themselves and the results that are expected of them.

Dealing with tension between autonomy and heteronomy: An issue for the organization

The *raison d'être* – i.e. the meaning – of activity is to respond to the insuffi-ciencies and the exteriority of prescriptions. Prescriptions are insufficient because the task is invariably overwhelmed by the real-world situations that it anticipates. It is also exterior, because it is based on reasons that are heteronomous to the acting subjects.

Therefore, ergonomics aims to design organizations capable of sup-porting autonomy, since one must be able to depart from prescriptions to carry out the strategic intention of the task. But in order to depart from the prescription, people must also be able to mobilize their own reasons for acting – i.e. their subjectivity. Therefore, ergonomics advocates an organi-zation where the company's reasons resonate with the people's reasons for acting – that is, an organization where heteronomy cannot be radical, and must instead be relative, and which accepts a degree of autonomy, which must also be relative since no activity can occur without it. The 'task' can only serve as a resource for the activity if it can combine the sources of prescription – those related to what the company expects of the worker, and those which the worker reintroduces himself or herself – his or her dispositions for action, updated in the face of the real world (Hubault and Sznelwar, 2011).

The means that agents have of experimenting, debating, designing efficient rules and making decisions depend on this fragile balance. All of this can be termed the deontic activity (Dejours, 2009).

Ergonomics supports the idea that the organization of work is a pro-cess in which the mobilization of subjective resources – the body, the subject – plays a central part.

In the Taylorian and Fordian system, the work organization, placed under the sole responsibility of engineers, relies on the hypothesis that work is identical, in its nature, to the operation of machines. This hypothesis is the basis of an organization that defines work in terms of *execution*. In the man-machine pairing, man forms with machines a system that is, in the end, purely technical (human engineering). Subjectivity – and, as a consequence, competence, cooperation, engagement and autonomy – has no place in this organization. There is therefore a certain degree of consistency between the absence, in the model of scientific organization of work, of work in any form, and the central role ascribed to engineers in managing the production system. With the evolutions of the basis for competition between firms and the new configurations of production that derive from this evolution, the drivers of performance are no longer just technical. The set of resources that must be mobilized no longer depend exclusively on engineering, but on the engagement, subjectivity and inter-subjectivity of agents in work situations that no longer require them to execute tasks and apply rules, but to respond to the unforeseen elements of these situations. From there, one must understand how value creation is driven by competence, cooperation, engagement and autonomy; understand that organizational innovation is no longer just a result of technological innovation alone. All this becomes a requirement in the work of managers, shifting the balance of their work toward the social and political sciences. By (re)focusing on the issues of work, these sciences have completely altered their relationship with organizations, and along with it, the role that ergonomics can play in these organizations.

Yet, it is more than ever difficult to satisfy this requirement. Management is finding it difficult to set itself apart from engineering, although it is in this difference that its specific character resides. It is the idea of *work as a resource* that is the source of this resistance, echoing the heritage of Taylor and Ford, which reduces activity down to a succession of operations. Resource and development are therefore two inter-related concepts. One cannot think about the management of human resources without thinking also about the development of this resource, both as a means and as a project in the process of value creation.

Being competent in *resource management* is therefore a key issue for organizations wishing to

- Transform the function of standards, abolishing their strict status as a heteronomous requirement (i.e. what must be and what one must comply to) in order to give them status as a resource and as a reference (thinking out what one can do, what can help in getting it done) that can be mobilized in situations that appeal to subjectivity
- Design organizations beyond an engineering approach (that is focused on the control of prescription through standards) highlighting the

role of management (a management that is mindful of developing immaterial resources – trust, cooperation, health – allowing the work to be done)
- Develop models for the assessment of work activity, which value not only observable and tangible modes of operation, but also individual and collective forms of engagement of subjectivity in one's activity: cooperation, responsibility, mutual aid (Clot, 2008; Dejours, 2009; du Tertre and Hubault, 2008)

Setting the organization: A relationship with risk

From engineers to managers, there is an evolution that is concerned with the relationship that companies have with risk. In the vision of engineers, that of a world that can be mastered, work is deployed in a universe that is controlled and known beforehand, through a set of rules whose application must be verified by management. In the vision of managers, the world is emergent and uncertain. Work is better recognized as a resource that aims to cope with a universe that is indeterminate and varying. Risk is contained within this indetermination and, as a consequence, within the impossibility of defining all responses beforehand. The existence of this risk and how it is taken on both constitute the very nature of work.

To claim this may seem imprudent, given the extent to which work is an object of regulations, technical responses and organizational intentions – all of which are intended to protect it from risk. But it would be just as reckless to attribute too much power to these provisions, since applying them strictly can easily turn into constraints and risks for those who must apply or enforce them. Let us take the example of an experienced worker using safety goggles while manufacturing complex and unique pieces. The goggles protect him from projections of filings and cutting oil. He knows this and wears them. However, he does not wear them all the time. This illustrates the shift from an initial rule to another rule, produced by autonomy: 'taking the goggles off' might well be viewed as an offence (when viewing the world as something that can be mastered), but also as a response (in a view of the world as an emerging system). Our worker will remove his goggles when, in a given situation of production, he opts against keeping them on. Indeed, he was requested to achieve a certain quality of production 'on the first go'. The need for added value in the pieces originates from a message, clearly perceived by all members of personnel, in a very competitive market. Failing to manufacture a piece correctly, having to do it again, implies disappointing a customer and ultimately endangering one's job. Therefore, our worker has elected to not use the goggles that, when splashed by the oil projections, might have prevented him from processing all of the information – notably visual – that is required for defining the machine settings and operating it. Here, working consists in being able to identify and locate these various kinds of risks in order to know how to act – that is, being able to assess what is going on, with respect to what might happen if one does not react suitably (Nascimento et al., this volume). Engaging this ability explicitly mobilizes subjectivity – that is, the sensitive ability to be affected by what is going on. Subjectivity, here, becomes a springboard and a resource for production. For our fitter, *being affected by what is going on* is related to the commitment of the company to its customer: it involves the brand image, the trust of the company, but also of the work collective and the worker himself. This introduction of subjective aspects in the realization of a task – manufacturing a piece to specifications – is part of the task itself, and management must pay attention to these aspects. Let us emphasize this point: the more the situations encountered by workers are ambiguous and enigmatic in relation to what they should be at the moment the rules are designed, the more it is necessary to mobilize the subjectivity of workers. The *presence* of management *at that time* is essential. Indeed, it will find it very difficult to understand an experience it has not shared – or to enhance its value, to support it, and make it evolve. Setting the level of proximity thus becomes a key strategic issue for management, in terms of both driving events and assessing the contribution of real work to results.

Let us now focus on evaluation. Every organizational system rests on an assessment of risk, of the means that are to be implemented to deal with these risks, and of the ways in which the real efficiency of the whole can be judged. The dominant trend today seems to rely on benchmarking and the promotion of 'best practices'. The problem is this: benchmarking rests on the hypothesis of a structural equivalence between events. This leads to supporting a strategy of purely incremental innovation, based on a logic of imitation and alignment (du Tertre and Hubault, 2008). This is the essential goal of Lean manufacturing, and the basis of the criticism addressed to it by followers of a clinical approach to work (Hubault, 2012). Benchmarking is particularly irrelevant in the case of service dynamics, for two key reasons, which should be emphasized here: (1) the immaterial resources that are mobilized are not measurable, and therefore lack any common ground for comparison, and (2) they can only be evidenced in the real-world situations that reveal them – trust, cooperation and skill can only be well assessed in a crisis, where they can save the situation. These reveal themselves in a timeline that does not rely on continuous time, as is the case in the valuation of capital, but on discontinuous, emerging time, which cannot easily be used as a grounds for comparison.

Essentially, benchmarking perpetuates the Taylorian model, which, by reducing activity to a system of operations and the productivity of work to the temporal costs of normalized, elementary operations, relies on integrating a finality to these operations. This vision is akin to neutralizing the issue of meaning (see the example of the quantity of grains on a car bumper) by reducing it down to the procedures, and confusing evaluation (e.g. making a trade-off between sanding down all the grains and postponing delivery, and abandoning sanding and being on time) and counting (e.g. counting the number of grains). Yet, the economic dynamics tend to focus increasingly on immaterial attributes. This implies that one can no longer just measure to perform an assessment, but should mobilize a value judgement.

Setting the organization: A relationship with the real world

Working means being concerned with all the forms of the real world that emerge in addition to what had been supposed, in order to cope with them and make use of them. Managing therefore means being concerned with events (both positive and negative) that make these forms emerge, and with the means that allow workers to respond to them. This plurality of the forms of the real world is truly *in* the real world. It is a given fact of the situation, which cannot necessarily be programmed, but can be foreseen. The analysis of 'real-life work' must be able to recognize work following this goal: to describe and understand what the worker does, but also what he cannot and does not do, what he is preventing, and what is preventing him from doing the job, what he is making come to pass, what he is seeking, what he might do, what he should be doing, etc. for the product, the customer, the company, the world, the collective, for himself, etc.

An 'organized work' is always proposed as a response to deal with the uncertainty to which real-life work will be confronted. But it also organizes an experience in which workers will put the limits of this response, as they experience these limits for themselves. This will inform them that this particular response is not a 'way out'. This experience leads workers at the heart of an uncertainty that constitutes, in the end, the stake of work itself. One might say that by choosing the risks that it decides to protect itself from, the organization decides, at the same time, the risks it shifts to the 'unregulated' space of real-life work.

Setting the organization: A specific relationship with subjectivity

Two contradictory logical approaches of a service – the service delivery and the service relationship – are present in real work situations. It is therefore in the activity itself that everything is sorted, through a dimension of activity that is not well known, and yet which has always been a component part of it: subjectivity.

Work is concerned just as much with the relationship of workers to the rule – as far as system operation is concerned – as with the relationship to values in the order of subjectivity. Real-life work always arbitrates between these two normative ranges. Hence, both the work organization and the management that drives it are confronted with a choice:

- *To justify the collective or to justify autonomy only in terms of their utility*: But then, how can one account for engagement and involvement if one only recognizes these to make use of them, with a strictly utilitarian – and therefore, heteronomous – view of perfecting their operation?
- *To justify the collective or to justify autonomy only in terms of subjective values*: But how can one understand, then, their effectiveness if one does not recognize their rationale in the very nature of the situations involved?

We can therefore see that it is a key issue to develop a view of the organization as something that mobilizes the resources necessary to managing the tension between these heterogeneous designs. Indeed, in the end, everything depends on the ways in which these situations respond to one another: the areas that the organization identifies as a risk – see, in our example above, the risk of oil projections in the eyes if one does not wear goggles – are not necessarily the areas that the worker identifies as a risk to himself (for the fitter, the risks which he identifies in relation to 'not getting it right the first time').

To achieve this vision, one must recognize the power of subjectivity in the battle with the 'unexpected elements of work' – that is, anything that will constitute an event and justifies the active presence of the subject. But to draw this line of reasoning to a close, one must be able not only to relate subjectivity with its economic utility, but also to understand why the subject has a stake in being able to do something for himself or herself out of what is happening (or happening to him or her).

The managerial stakes of autonomy

Ergonomics advocates a view of human resources management that is geared not toward managing people, but toward managing the productive power of their activities. This orientation is concerned with the way in which the organization constructs a relationship with autonomy: on the one hand, the autonomy of people in their work, and on the other hand, the autonomy of work in the processes of value creation. Indeed, there is no way of truly supporting autonomy if one grounds its development only in what the organization expects out of it.

When this orientation is not clear, this leads to an *injunction* of autonomy – which is increasingly prevalent today, and increasingly pathogenic. This may, for example, take on the form of the modern *obligation* to promote collective work. Because it is impossible to prescribe the task with enough precision, the order to work is increasingly addressed to the collective in modern organizations. However, such organizations continue to apply the form of control that is established in relation to an individual task; the development of *individual performance evaluation* is a very clear illustration of this.

In other words, the issue is to coordinate two dynamics: one that sets the need for autonomy, and the other that reaffirms the need for a central control – and contradicting the first. The issue is to recognize this contradiction in order to manage it. The wish is either to maintain centralized control over process regulation, or to develop a process-based vision, the only vision that can offer an opportunity for autonomy. Of course, these two modes of regulation are intertwined in the activity of every person. However, confusing the two modes leads to the impossibility of understanding autonomous regulation.

The heart of the matter therefore boils down to this: management is structurally set in a logic of defiance related to the people it deals with. It has increasingly, itself, fallen prey to defiance on the part of the upper layers of decision-making, and so on, moving further and further upward. This position is practically a complete reversal of the hypothesis of a human resource, which the organization could admit to being able to count on. This failure to consider work as a resource testifies to a general mode of thought that constantly wagers against the real world that escapes it, against the ability to face risk, against work, and against autonomy. It is therefore the very idea of a 'process' that cannot take hold, and the very idea of development that is prohibited. The key features of Taylorian work can be found here: it is closed onto itself, it is pure 'operation', and it leads to no development, be it subjective or economical; nothing can be built onto it. This is why it has an alienating character: activity can only lead to servitude if it leads to perpetuating its initial stance.

What about the new forms of work organization?

The idea of organizational change emerged in the 1980s and has been a strong focus of attention in companies. Companies rely on organizational change, occasionally, regularly, gradually or suddenly, depending on the requirements of increased competitiveness, on certifications demanded by customers, or simply on the introduction of new technologies. Every time, these changes strive toward organizations of work that are less fixated upon standardization of work and heteronomy.

The Valeyre inquiry on conditions of work in the European Union in 2006 proposes a categorization in which two forms of work organization might correspond to this goal: Lean production and learning organizations. Lean production is characterized by a strong tendency toward the diffusion of work in teams, task rotations, versatility and quality management through strict standards and self-monitoring conditions. Autonomy is described as relative (*controlled*) in those situations, although these conditions are exposed to unexpected problems in the form of time constraints. *Learning organizations* are related to Lean production (teamwork, self-monitoring, unexpected situations) but differ from it by allowing a greater autonomy at work and a lesser time constraint. It seems that these two principles are promoted as means to cope with complex, nonrepetitive situations. Although the results of the inquiry show a good correlation between learning organizations and good working conditions, Lean production, in contrast, is very close to – or indeed, is worse off than – Taylorian organizations.

Activity ergonomics is less familiar with learning organizations than it is with Lean production, and knows how to analyze the latter's low scores (Bourgeois, 2012; Hubault, 2012). In intent, Lean production advocates departing from Fordian modes of prescription. For example, it is requested that the prescribing agent should not ignore the realities of the field, and should be ready to listen to the workers. However, Lean production does not give up on standardization; on the contrary, standardization remains its horizon. Indeed, Lean production wagers on the capacity of standardization for improvement – which it aims, precisely, to develop – by taking into account the difficulties encountered at work, notably during kaizen workshops. Although Lean production does request the word of the worker, which is a key difference with Taylor's response to workman Shartle, this approach finds it difficult not to instrumentalize subjectivity, all the while recognizing this subjectivity. This can be seen in very concrete terms, by the way in which a gesture, a movement or outstanding parts continue to be defined as 'waste'. Or in how the feedback of workers who are requested to express themselves in kaizen workshops is restricted to the topic of the difficulties encountered, and does not talk about the experience of action. From this point of view, Lean production is not very different from Taylorism. It can even be viewed as a more arduous experience, because the whole system is supported, both ideologically and methodologically, by a promise to listen to the worker.

This chapter has sought to account for the resource that activity represents, as well as to describe the reasons and the issues why it is largely ignored in the current evolutions of organizations. However, it would be fairer to view the problem in terms of the fear that the integration of human activity awakens in these organizations. What can we do, then? Certainly, one should improve the demonstration that taking a chance is better than denying the issue. This suggests the need for a closer collaboration of ergonomics with management science, economics, psychology and sociology, in order to design evaluation systems that include concepts of autonomy, subjectivity, leeway and regulation, and are compatible with managerial concerns. This is one of the current pathways for the development of activity ergonomics.

References

Béguin, P. (2007). Taking activity into account during the design process. *@activités*, 4(2), 107–114. Retrieved from http://www.activites.org/v4n2/beguin-EN.pdf.

Bourgeois, F. (2012). Que fait l'ergonomie que le Lean ne sait/ne veut pas voir? *@ctivités*, 9(2), 138–147. Retrieved from http://www.activites.org/v9n2/v9n2.pdf

Clot, Y. (2008). *Travail et pouvoir d'agir*. Paris: PUF.

Daniellou, F. (2004). L'ergonomie dans la conduite de projets de conception de systèmes de travail. In P. Falzon (Ed.), *Ergonomie* (pp. 359–373). Paris: PUF.

de Terssac, G. (2012). La théorie de la régulation sociale: reperes introductifs. *Revue Interventions Économiques*, 45, 1–16.

Dejours, C. (2009). *Travail vivant*. Paris: Payot and Rivages.

du Tertre, C. (2007). Création de valeur et accumulation: capital et patrimoine. *Economie Appliquée*, 3, 157–176.

du Tertre, C., and Hubault, F. (2008). Le travail d'évaluation. In F. Hubault (Ed.), *Évaluation du travail, travail d'évaluation* (pp. 95–114). Toulouse: Octarès.

Friedmann, G. (1956). *Le travail en miettes*. Paris: Gallimard.

Hubault, F. (2004). La ressource du risque. In F. Hubault (Ed.), *Travailler, une expérience quotidienne du risque? Proceedings of the Paris 1 Symposium* (pp. 207–220). Toulouse: Octarès.

Hubault, F. (2012). Que faire du Lean? Le point de vue de l'activité. Introduction. *@ctivités*, 9(2), 134–137. Retrieved from http://www.activites.org/v9n2/v9n2.pdf.

Hubault, F., and Sznelwar, L. I. (2011, April). Psychosocial risks: when subjectivity breaks into the organization. Presented at Proceedings of ODAM, Grahamstown, South Africa.

Laville, A., Teiger, C., and Duraffourg, J. (1972). *Conséquence du travail répétitif sous cadence sur la santé des travailleurs et les accidents*. Report 29, Ministry of Education. Paris: Occupational Physiology and Ergonomics Laboratory, CNAM.

Maggi, B. (1996). La régulation du processus d'action de travail. In P. Cazamian, F. Hubault, and M. Noulin (Eds.), *Traité d'ergonomie* (pp. 637–659). Toulouse: Octarès.

Teiger, C. (2008). Entrevue guidée avec Hélène David et Esther Cloutier, Défricheurs de pistes. *Pistes*, 10(1). Retrieved from http://www.pistes.uqam.ca/v10n1/articles/v10n1a4.htm.

Valeyre, A. (2006). *Conditions de travail et santé au travail des salariés de l'Union Européenne: des situations contrastées selon les formes d'organisation*. Working Paper 73. Noisy-le-Grand, France: Centre d'Etudes de l'Emploi.

Van Belleghem, L. (2012, September). Simulation organisationnelle: innovation ergonomique pour innovation sociale. Presented at Proceedings of the 42nd SELF Conference, Lyon, France.

Vicente, K. J. (1999). *Cognitive work analysis: toward safe productive and healthy computer-based works*. London: Lawrence Erlbaum Associates.

chapter seven

Constructing safety
From the normative to the adaptive view

Adelaide Nascimento, Lucie Cuvelier, Vanina Mollo,
Alexandre Dicioccio and Pierre Falzon

Contents

This chapter describes various faces of safety and questions the conditions of development of a form of safety that includes the co-design of rules, their appropriation, their use and the adaptations that are necessary in real-life situations. The literature in the social sciences – ergonomics, sociology, psychology, etc. – has emphasized the complexity of this issue, which cannot be simply reduced to a static articulation between, on the one hand, the production of formal rules, and on the other hand, the

95

use that is made – or not made – of these rules (Amalberti, 2007; Bourrier, 2011; Daniellou, 2012; Dien, 1998). It is useful, at this point, to take into account the adaptive, dynamic and developmental character of safety, in order to continue to make progress in this field.

Our work is located within this constructive view of safety. Safety does not result merely from eliminating malfunctions – i.e. suppressing technical and organizational variability – or solely from defining pre-planned responses to errors and failures (via standardization of human action). It is the result of the ability to succeed in varying conditions, making use of all available resources.

However, it is not possible to 'let everyone do as they please', because situations may be very hazardous. Nor is it possible to forbid any adaptation – because variability does exist, and the blind application of rules could lead to suboptimal decisions with respect to specific criteria. We must therefore allow for both a reasonable loosening of the rules, and processes for managing it in an appropriate way – that is, by defining places where the trade-offs that are made can be discussed. These elements of activity are closely connected with developmental processes. Safety is constructed as skills are developed, which creates new resources for acting safely. It is in this sense that we will talk about 'constructed safety' in this chapter.

Models of safety are a hot topic of debate today. Because of this, one can note the existence of many terms, some of which relate to concepts that are quite similar. This chapter will attempt to clarify the equivalences, resemblances and differences between these terms.

Regulated safety and managed safety: From the addition to the articulation of safeties

Regulated safety and managed safety

Since the past few years, two fundamental paths have been identified to achieve safety (Amalberti, 2007; Daniellou et al., 2011; Falzon, 2011; Pariès and Vignes, 2007): *regulated safety* and *adaptive safety*.

Regulated safety aims to control risks by regulating work practices. It relies on formulating rules (procedures, frameworks, prescriptions, etc.), on diffusing these rules amongst stakeholders and on enforcing their application (Hollnagel, 2004, 2006). These rules have various origins: models of the operation of a technical system, standardized empirical data derived from research and experience feedback of incidental or accidental situations. Designers aim to anticipate as many situations as possible, to ensure that the operator will not have to construct a solution on the spot. Regulated safety can therefore constitute a resource for action, in the sense that it provides a framework for action that often encompasses situations that have been encountered in the past in order to prevent situations that may occur in the future (Mollo, 2004).

Conversely, managed safety relies on the capacity of operators for initiative, either alone or as a group, when dealing with unforeseen events and with the natural variability of the real world. This approach derives from the idea that it is pointless to believe that everything can be foreseen – human intervention is therefore necessary to ensure reliability. As Reason et al. (1998) point out, 'There will always be situations that are either not covered by the rules or in which the rules are locally inapplicable'.

Some authors (Cuvelier and Falzon, 2010; Dien, 2011; Pariès, 2011) have noted that regulated safety and managed safety were initially described, either as a sum of added terms or as two incompatible approaches that were contrary to another; the extension of the field of regulated safety and of foresight gives greater precedence to formal rules, reducing the autonomy of agents: too much regulated safety kills managed safety (Morel et al., 2008; Daniellou et al., 2011; Nascimento, 2009).

It seems to us that the question is not so much one of setting a 'cursor' between the domain of regulated safety and that of managed safety – but one of producing relevant rules and determining how these rules are used, transformed and invented reasonably in real-life situations.

Safety in action and effective safety

The concept of 'safety in action' takes a step in this direction. It is presented not so much as a dichotomy than as a combination between rules and the real-time management of the situation. It is 'the way in which subjects go about acting safely in the face of disruptions, and about managing their own actions which are not always optimal with respect to the rules' (de Terssac and Gaillard, 2009, p. 14). The authors argue that safety in action is arbitrated by professionals themselves, depending on the situation, complementing formal rules or contradicting them. In this sense, professionals combine rules *in action*, electing to use them or not – and above all, inventing new rules in order to act safely (de Terssac and Gaillard, 2009, p. 14). It is a manifestation of safety in the here and now.

Following this view, safety rules and rules of action cannot be considered separate from one another. The management of safety should be viewed as 'an action that is attached to professional action and flows together with it, and not as an action that is detached, separate, and different from professional action' (de Terssac and Gaillard, 2009, p. 16, our translation).

More recently, and in a sociological perspective, de Terssac and Mignard (2011) made use of the dynamic nature of the life of the rules in their analysis of the catastrophe at the AZF chemical plant. They introduced the concept of effective safety to complement that of safety in action. Effective safety is viewed as a process during which there are stages of flexible safety, imposed safety and negotiated safety. It is 'the way in which subjects shift from a safety that is set by rules to safety in action, through the transformation of formal rules into shared obligations which every person commits to upholding: engagement, appropriation, understanding, and knowledge-based coordination form a set of social rules that are invented and mobilized to "act safely"' (our translation).

The normative approach and the adaptive approach

As we have seen above, the debate as it currently stands focuses on how to articulate two pathways to achieve safety – one that is anticipated, predetermined, reactive or 'top-down' and imposed to the sharp-end operators, and the other that is reactive, adaptive and based upon individual and collective competence (Pariès, 2011). In this section, we will describe these two positions.

The first position emphasizes regulated safety and ignores – or pretends to ignore – adaptive safety. In this view, any manifestation of adaptive safety constitutes a failure of the system and should be prohibited. Safety is sought through conformity. It is possible to assess the level of safety by assessing the level of conformity to prescriptions. This position will, in the remainder of this chapter, be termed the *normative approach*.

The second approach has been described as *adaptive* (Falzon, 2011). It relies on all of the resources that are available – that is, on the one hand, the procedures, rules and standards that are edicted by organizations and tutelary authorities, and on the other hand, the rules that are constructed locally, the ad hoc procedures that are constructed to cope with the variability of real-world situations. The adaptive approach is interested in safety in action (de Terssac and Gaillard, 2009) and relies on the intelligence of the agents involved in the situation to act safely. It therefore combines regulated safety and managed safety, making it possible to ensure system resilience – that is, 'the intrinsic ability of a system to adjust its functioning prior to, during, or following changes and disturbances, so that it can sustain required operations under both expected and unexpected conditions' (Hollnagel, 2010).

This distinction between normative and adaptive safety can also be found in the two models of application of rules described by Dekker (2003), who draws an opposition between 'blind adherence' to rules and the application of rules as a complex cognitive activity – an activity that is 'substantive' and skilful.

The first model rests on four postulates:

1. Rules constitute the safest way to perform a job. This characteristic echoes the principle of the 'one best way'.
2. Adherence to rules then consists in a mental activity that is mainly based upon simple rules of the 'if ... then' type.
3. Safety results from operators conforming to procedures.
4. To improve safety, organizations must make an investment to ensure that workers know the procedures, and that they comply with these procedures.

In this model, the emphasis is placed on conformity to rules, rather than on the relevance of the response to new or unforeseen situations. This may undermine the capacity of an organization for resilience (Daniellou et al., 2010).

Conversely, the second model posits that rules, although they constitute resources for workers, are not enough to cope with every single work situation. This model is, according to Dekker (2003), based upon the following four assertions:

1. Rules are resources for action. They do not specify each and every one of the circumstances in which they apply, and therefore cannot dictate their own application. They cannot ensure safety on their own.
2. The correct application of procedures to situations can be an activity based upon competence and experience, i.e. be substantive and skilful.
3. Safety results from the ability of operators to judge when and how they should (or should not) adapt procedures to local circumstances.
4. To improve safety, organizations must control and understand the reasons underlying the discrepancies between rules and practice, and develop the means to help workers to judge when and how to adapt these rules.

Daniellou (2012) and Dien (2011) point out that the attitude of management is a fundamental factor to make some progress in the operational connections that move an organization toward an adaptive approach. It often emerges that this attitude is neither consistent nor homogeneous, because it is overly dependent on the success or failure dimension of the attempted action. Before going into detail on this aspect, it seems useful to draw some distinctions between the various forms of safety in action, concerning workers at the sharp end. To do this, we draw support from four empirical studies in the fields of medicine and aeronautics (Cuvelier, 2011; Dicioccio, 2012; Mollo, 2004; Nascimento, 2009).

The adaptive approach and forms of safety in action

When faced with the diversity of real-world situations – whether this diversity is more or less planned, and whether it is routine or exceptional – the forms in which regulated safety and managed safety can be articulated with one another may vary. Two possibilities can be distinguished:

- The first possibility is that there are no formal rules that are precisely suited to dealing with the situation, either because the existing rules are too generic (Section 8.3.1) or because some cases are not covered by the rule (Section 8.3.2). Managed safety then enters the fray to support regulated safety, to complement formal rules. Following this view, workers do not have to make any trade-off regarding whether or not to follow the rule, since this rule does not really exist. They must therefore invent new rules, possibly drawing inspiration from existing rules.
- The second possibility is that formal rules do exist, but their application is being questioned. This may occur when the context that had initially been foreseen is not present (Section 8.3.3) or when applying the rule seems counterproductive (Section 8.3.4). This time, managed safety steps in, to decide whether or not to make use of the rule (concept of possible violation of the rule) and to decide what degree of transgression of regulated safety is acceptable.

Rules are generic and must be particularized

In some cases, the variability and complexity of situations are such that safety rules cannot prescribe in detail the ideal conduct that workers should adhere to. The rules indicate criteria, properties and possible means to reach safety goals, but the implementation of these rules always requires interpretations and decisions that take into account the singular character of each situation. Illustrations of this combination can be found in the medical field.

In this field, rules (e.g. recommendations, treatment protocols, benchmarks for best practices) are grounded into evidence-based medicine (EBM). Several research projects have allowed us to characterize EBM as 'a relatively flexible organization' that does not prevent the development of expertise or autonomy in decision-making (Cuvelier et al., 2012; Mollo and Falzon, 2008). For example, one can see that while they adhere to the rules in accordance to EBM, anesthetists can propose different strategies for an intervention (variability between anesthetists working on the same medical case), even when working on frequent cases that are deemed easy or usual (Cuvelier et al., 2012). Thus, compliance with the rules does not always eliminate the variability of possible solutions and, because of this, does not systematically lead to identical strategies for action. In other words, regulated safety does not 'crush', in this case, the situated decisions made by the subjects, and it always requires managed safety as a complement. For instance, in oncology, therapeutic benchmarks contain a number of decision criteria whose values are adapted when they are combined with the specific characteristics of the patients to be treated. In particular, the patient's age is expressed in chronological terms, whereas doctors reason from the patient's physiological age, and may then end up tailoring the treatment that is recommended by the benchmark.

Whereas in the normative approach, this variability of possible solutions might be seen as regretful, the adaptive approach views, instead, rules as resources that support and provide input for managed safety, without substituting themselves for managed safety. In this case, regulated safety and managed safety constantly combine with one another, and both of them are necessary to achieve overall safety. Indeed, it can be noted that this variability of work practices does not make it impossible to reach a high level of overall safety. On the contrary, various works have shown that the diversity of strategies allowed by the rules can, in some conditions, prove to be a margin of adaptation that is profitable for the development of safety (Amalberti, 1992; Cuvelier et al., 2012). Returning once again to our example, anaesthetic medicine is an 'ultrasafe' system that is viewed as pioneering the field of healthcare safety (Amalberti et al., 2005). Analyzing the activity of anaesthetists shows that choosing a strategy from the set of 'possible strategies' results from a subtle trade-off, allowing each worker to find the most appropriate balance between the singular features of the case being treated, on the one hand (the patient, the patient's decisions, the family, the complexity of pathologies, organizational constraints, etc.), and the resources that are available to deal with this case, on the other hand (the doctor's own skill set, the skills of the medical team, the mastery of equipment and medical techniques, the development of rules of the trade, etc.). It is this process of matching the features of the case with the available resources, a process that is always singular, that allows each doctor and each medical team to work within their respective fields of expertise, and therefore to reach a very high level of safety – regardless of the difficulty, the rarity or the unpredictability of a particular case.

Rules may not cover some of the cases

Some fields of human activity are less prescriptive than others, in terms of well-described work procedures. In these situations where there is no external rule, the standard, the reference can be a priori internal to an individual worker or to a trade. Hence, the activity of workers will essentially depend on their ability to 'invent' new rules. Yet, when the characteristics of a given situation question the applicability of prescribed rules, these rules are not necessarily abandoned. The workers can reason from rules that might be applicable in the absence of any 'deviant' characteristics. For example, in oncology, therapeutic guidelines do not take into account some of the patients' medical characteristics (medical history, comorbidities, etc.). Similarly, there are no guidelines for the treatment of breast cancer in men. Yet, doctors reason from the guidelines that would apply if the characteristics we have just mentioned were absent, to determine the therapeutic strategy that is best suited to a particular case (since the guidelines guarantee the effectiveness of the treatment). In radiotherapy, when faced with a nonnominal situation for which there is no formal rule available to guide action (e.g. a breakdown of imaging equipment, lack of any accessories to position the patient correctly, etc.), professionals rely on their experience in the profession or in the department to solve problems. Thus, they construct metarules for action, based upon their knowledge of the patient in question and of the current stage of the treatment, of the behaviour of colleagues, of the availability of resources, and of the wish to develop a strong safety culture.

Be that as it may, it can be noted that the rules that are created rely on the entire set of available resources: existing formal rules and rules derived from individual or collective experience. These rules can become stabilized following the repeated occurrence of similar situations.

Technical operation is deficient or the rules are not adhered to by all workers

Managed safety also steps in when the context that was originally foreseen for regulated safety is not present. This is the case, notably, for technical failures (for example, equipment breakdowns) or in the case of cascades of rule violations. In radiotherapy, running late leads to violations of the control rules prescribed by the organization: patient files are filled in in a hurry, and successive controls are performed too quickly, or even not at all. For example, radiotherapists may decide not to validate the treatments they prescribe when these treatments match the protocol very closely and are therefore viewed as 'simple', or when there is no time to do so. To gain some time, a radiotherapist will only validate the treatment during the first weekly consult with the patient, i.e. 1 week after the beginning of the treatment. This nonnominal situation leads to a sequence of actions of recovery on the part of the x-ray operators, seeking to return to a nominal situation where the rule may be applied. The x-ray operators then try to contact the doctors to request the validation of their file. If this request is successful, then safety in action has made it possible to return to a normal situation. Hence, managed safety steps in to support regulated safety.

Applying the rule is viewed as counterproductive

However, in this same example, recovery may prove to be impossible because of the poor availability of doctors. X-ray operators must then perform a trade-off between treating and not treating a patient for whom the medical validation that is required by regulations has not been provided: to carry out the treatment and violate the procedure or to cancel the session and comply with the rule. The strategies of the x-ray operators are based upon a cost–benefit analysis of whether to violate or comply with the rule. This may lead them to make a reasoned trade-off and to apply the rule (and not carry out the treatment) or to violate it (carrying out the treatment without validation). Risks are present regardless of the decision made. In the first case, the patient will not receive the daily dose of treatment, and his or her chances of survival may be reduced. In the second case, going ahead with the session with a nonvalidated patient file leads to a risk of the treatment not being in accordance with regulations – which operators view as a less serious risk than the patient failing to receive the daily dose.

A second example is derived from the field of air transportation (Dicioccio, 2012). This field is a second reference to ultrasafe systems: all of the stakeholders involved must comply with an impressive number of safety rules and procedures. In this hyperprocedural system, particularly in aircraft maintenance, the adaptations that are made by sharp-end operators to cope with unforeseen events may lead to a failure to comply with rules to 'achieve expected performance' (Amalberti, 2007). At this point, managed safety steps in to make a trade-off between making use and not making use of existing rules, in order to preserve performance with an acceptable level of safety. When faced with an overabundance of rules and procedures, the technicians must adapt to the variations related to contextual factors and to the related constraints. Because of this, trade-offs are an expression of a reasoned adaptation of procedures by the workers (Amalberti, 2007). For example, in the event of identifying a defect in an aircraft that is set to take off, the technicians assess the severity of the defect and its possible evolutions, by carrying out a technical analysis of the defective equipment or associated systems, and of its impact on the running of the aircraft. Therefore, the risk taken in the trade-off – to allow or fail to allow the plane to take off – is minimized. The adaptation produced by the more experienced technicians manifests itself in the form of strategies of anticipation, supervision and implementation of means to reassess the situation. In this way, managed safety allows the technicians not just to decide what degree of transgression of the rules that they deem acceptable, but also to construct a strategy of action that is acceptable in terms of safety and performance.

In these examples, managed safety does not breed lack of safety, but acceptable risk: high performance is sustained with no adverse impact on safety.

The construction of safety

Regulated safety and managed safety are therefore constantly intertwined within the workers themselves, but also at the level of organizations (reconstruction of rules and supervision by managers). We will make a distinction between four types of organizational conditions that foster the construction of safety, where operators and managers develop shared knowledge regarding issues of safety:

- Processes of 'integrated' design of rules that link together top-down and bottom-up processes
- Taking into account the collective aspects of work in organizations and training programs, allowing each person to know about the activity of others, and to take this knowledge into account in his or her own activity
- The dynamics of collective decision-making, which make it possible to discuss the situations encountered and to construct a common frame of reference
- A mode of management, which places a premium on understanding the decisions made by agents rather than on calling them to order, and on autonomy and responsible behaviour on the part of stakeholders, all of whom are seeking shared and reasoned trade-offs

The integrated design of rules

The categorization scheme and the examples presented above consider only the users/adapters/inventors of rules at the sharp end. Yet, system safety does not rely solely on field workers, as has been shown in the analysis of various disasters (Chernobyl, Three Mile Island, the explosion of the BP refinery in Texas City, etc.) One cannot ignore the role played by the designers of formal rules and by management. These are heavyweight stakeholders in the construction of safety.

An approach grounded in the integrated design of rules implies that the rules are not just learned and applied, but also understood by the workers. This means, on the one hand, that the rules must be grounded in an analysis of work activity, and on the other hand, that 'the rationale that these rules are based on' and their 'organization into a system that is consistent for action' are all handed down to the workers they are intended for (Mayen and Savoyant, 1999, our translation). Hence, designing rules in an integrated manner is one condition for the 'intelligent use' of procedures that, in particular, allows discrepancies when the rules do not 'fit well with the real world' (Dien, 1998). This supposes that the designers of the rules have access to genuine knowledge of system operation and of the work that is carried out by operators. In other words, the point of view of system designers and that of the workers should be compatible, and the relevance of rules will depend on their ability to take into account the safety practices developed in the field (de Terssac and Gaillard, 2009; Dien, 1998).

The work of all in that of every worker

Knowledge of the work of one's colleagues is a fundamental piece of information for decision-making in high-risk situations and, in the end, for making work practices safer. In the medical field, this aspect is all the more crucial, as the care of a patient often relies on the coordinated involvement of multiple specialists. Although the immediate goals of these workers differ from one another, these specialists must reason on the basis of the knowledge regarding the actions that are possible for their colleagues, in order to manage potential interferences between the goals of different specialists (Mollo, 2004). For example, in radiotherapy, x-ray operators take into account the work habits of their colleagues to make decisions that aim to preserve productivity and safety at the sharp end (Nascimento and Falzon, 2009). Similarly, still in radiotherapy, medical physicists who are responsible for preparing a treatment take into account the work that is done at the sharp end when choosing a technique that will be complicated or hazardous to implement for the x-ray operators (Nascimento and Falzon, 2012). Similarly, in anesthetic medicine, doctors build intervention strategies that take into account their own resources (skills, attention load, instruments, etc.) as well as the resources of their colleagues (skills, expertise and preference of colleagues) and those of the work collective (rules of the trade, shared practices within a department) (Cuvelier et al., 2012).

An essential issue in the construction of safety is therefore not just to support existing collective safety practices, but also to develop innovative means for training, such as simulation-based training for reflective practices (Cuvelier and Falzon, 2010) or the implementation of spaces to debate work practices (Mollo and Nascimento, this volume). Collective decision-making meetings are one step in this direction.

Collective decision-making

One of the tools that is currently being developed to deal with situations that escape the rules is *collective decision-making*. The expected goal of this kind of decision is to improve the quality and reliability of decisions and to alleviate variations in practice*. Multidisciplinary consult meetings implemented in oncology are an example of this (Mollo, 2004). These meetings bring together experts from various specialties (surgery, oncology, radiotherapy, etc.) to propose solutions to situations for which a strict application of the therapeutic guidelines is either impossible or unsatisfactory. A study of how these meetings operated allowed us to show that beyond their primary function of assisting decision-making, they allow the development of shared rules at the local level, ensuring the sustained reliability of decision-making.

Discussing each situation also makes it possible to define, for the situations that escape the rules, the space of acceptable and unacceptable practices. In so doing, multidisciplinary consult meetings play the part of a safeguard, in the sense that they manage the situations that escape rules by ensuring sustainability and the development of reliability in decisions (Mollo and Falzon, 2008). However, the goal is more to define the space of acceptable and unacceptable solutions than to strive toward a single acceptable solution (although this may prove to be necessary in some situations). Thus, collective decision-making makes it possible to develop a reliable framework of reference, within which doctors are free to apply their own expert rules, while being assured of the reliability and collective acceptability of their decisions. Indeed, during these exchanges, doctors integrate some new knowledge that they would not have taken into account if they had reasoned on their own (Mollo, 2004).

Modes of management that are open to an adaptive approach

These are modes of management that place a high priority on understanding the decisions of agents rather than on calling them to order, and that provide some space for the autonomy and responsible behaviour of agents, who are all involved in the search for reasoned and shared trade-offs. Following this view, safety not only belongs to managers, but is the business of every worker. It relies on a balanced connection between top-down and bottom-up processes, in order to achieve an 'integrated' safety culture that relies on the confrontation of points of view and on a debate between management and operators (Daniellou, 2012).

High-risk organizations are confronted with complex activities, bearing varying events with high economic and safety stakes, related to multiple trade-off decisions made by workers at the sharp end. This suggests

---

* We will focus here only on collective situations of decision-making that involve equal status of the stakeholders present, or recognition of the interest of each of these stakeholders' expertise. The decisions dictated by the member or members of the collective with the highest status will not be considered here.

that managed safety rests on some kind of tacit agreement within the collective, hidden from the eyes of managers. However, management bears the heavy load of the trade-offs carried out by its workers to maintain performance and safety. This responsibility cannot be diluted in ignorance. Workers at the sharp end make trade-offs regarding making or not making use of the rules. Managers have many ways of dealing with such transgressions. Likening these transgressions to violations by punishing the perpetrators cannot ensure the primacy of regulated safety, and would make these transgressions clandestine (for fear of subsequent punishment). Pretending to ignore the truth would lead to a loss of visibility on the expressions of managed safety, in such a way that the margins of managed safety diminish over time. Yet, not doing anything or doing everything to avoid any problems is indeed a reality that has been observed in the past in both individuals and organizations alike (Gilbert, 2005).

However, there are other ways to manage safety and its development. For example, in a study carried out in the ultrasafe field of aircraft maintenance (Dicioccio, 2012), management bears a shared and consistent judgement on the true state of managed safety. For these managers, to interrupt and to resume the running of an aircraft with a minor defect are both acceptable decisions. In this particular case, the deciding factor is not the adherence to a rule, but rather the understanding of the (operational, technical and human) environment that has led to making the decision. This fact is a trace of the existence of a reasoned, shared culture of trade-offs. What the management says and does follows the same route as what is recommended to the sharp-end operators. Management not only has to ensure compliance with rules, but also must take part in the articulation between regulated safety and managed safety (Daniellou et al., 2010). In this sense, managers, by working on the articulation between these two forms of safety by the workers, foster a culture of reasoned arbitrage.

In these high-risk systems, accepting and acknowledging the existence of safety trade-offs in production, by giving back some autonomy to the workers (Amalberti, 2007), is one path toward ultrasafety. However, an attitude of openness and professional respect in relation to the trade-offs made by others finds its true meaning only as far as these workers remain within a safe space of operation. Sustained performance is not achieved at the expense of safety. Performance must be reasoned, when it is connected with safety. This implies that the trade-offs in favour of safety and performance should be the focus of experience feedback. Such feedback should be shared by all workers, including supervisors.

The question then becomes how to support such feedback on experience. Mollo and Nascimento (this volume) propose several methods to achieve this. In particular, a method for the confrontation of practices, termed differential judgement of acceptability of discrepancies (Nascimento, 2009; Mollo and Nascimento, this volume), can provide a framework to define the contours of a space of acceptable practices to replace the 'rigid line to be followed'. The exchanges between workers implementing this method would likely help construct a shared framework of reference, by tracing together the path for the expression of managed safety.

References

Amalberti, R. (1992). Safety in process-control: an operator-centred point of view. *Reliability Engineering and System Safety*, 38, 99–108.

Amalberti, R. (2007). Ultrasécurité, une épée de Damoclès pour les hautes technologies. *Dossiers de la Recherche*, 26, 74–81.

Amalberti, R., Auroy, Y., Berwick, D., and Barach, P. (2005). Five system barriers to achieving ultrasafe health care. *Annals of Internal Medicine*, 142(9), 756–764.

Bourrier, M. (2001). *Organiser la fiabilité*. Paris: L'Harmattan.

Cuvelier, C. (2011). De la gestion des risques à la gestion des ressources de l'activité – Étude de la résilience en anesthésie pédiatrique. Unpublished doctoral dissertation, CNAM, Paris.

Cuvelier, L., and Falzon, P. (2010). Coping with uncertainty: resilient decisions in anaesthesia. In E. Hollnagel, J. Pariès, D. Woods, and J. Wreathall (Eds.), *Resilience engineering in practice: a guidebook* (pp. 29–43). Aldershot, UK: Ashgate.

Cuvelier, L., Falzon, P., Granry, J. C., Moll, M. C., and Orliaguet, G. (2012). Planning safe anesthesia: the role of collective resources management. *International Journal of Risk and Safety in Medicine*, 24, 125–136.

Daniellou, F. (2012). *Facteurs humains et organisationnels de la sécurité industrielle. Des questions pour progresser*. Toulouse: FonCSI.

Daniellou, F., Simard, M., and Boissières, I. (2011). Human and organizational factors of safety, state of the art. *Cahiers de la sécurité industrielle 2011-01*, Toulouse: FONCSI. www.icsi-eu.org.

de Terssac, G., and Gaillard, I. (2009). Règle et sécurité: partir des pratiques pour définir les règles. In G. de Terssac, I. Boissières, and I. Gaillard (Eds.), *La securite en action* (pp. 13–34). Toulouse: Octarès.

de Terssac, G., and Mignard, J. (2011). *Les paradoxes de la sécurité. Le cas d'AZF*. Paris: PUF.

Dekker, S. (2003). Failure to adapt or adaptations that fail: contrasting models on procedures and safety. *Applied Ergonomics*, 34, 233–238.

Dicioccio, A. (2012). Articuler sécurité et performance: les décisions d'arbitrage dans le risque en aéronautique. Unpublished doctoral dissertation, CNAM, Paris.

Dien, Y. (1998). Safety and application of procedures, or how do "they" have to use operating procedures in nuclear power plants? *Safety Science*, 29(3), 179–187.

Dien, Y. (2011). Sécurité réglée, sécurité gérée: une problématique à redéfinir? Presented at Les entretiens du Risques 2011, Institut de Maîtrise des Risques, Gentilly.

Falzon, P. (2011). Rule-based safety vs adaptive safety: an articulation issue. Presented at 3rd International Conference on Health Care Systems, Ergonomics and Patient Safety (HEPS), Oviedo, Spain.

Gilbert, C. (2005). Erreurs, défaillances, vulnérabilités: vers de nouvelles conceptions de la sécurité? In *Cahiers du GIS Risques collectifs et situations de crise. Risques, crises et incertitudes: pour une analyse critique*. Grenoble: MSH-ALPES.

Hollnagel, E. (2004). *Barriers and accident prevention*. Aldershot, UK: Ashgate.

Hollnagel, E. (2010). Prologue: the scope of resilience engineering. In E. Hollnagel, J. Pariès, D. Woods, and J. Wreathall (Eds.), *Resilience engineering in practice: a guidebook*. Aldershot, UK: Ashgate.

Hollnagel, E., Woods, D., and Leveson, N. (2006). *Resilience engineering: concepts and precepts*. Aldershot, UK: Ashgate.

Mayen, P., and Savoyant, A. (1999). Application de procédures et compétences. *Formation Emploi*, 67, 226–232.

Mollo, V. (2004). Usage des ressources, adaptation des savoirs et gestion de l'autonomie dans la décision thérapeutique. Unpublished doctoral dissertation, CNAM, Paris.

Mollo, V., and Falzon, P. (2008). The development of collective reliability: a study of therapeutic decision-making. *Theoretical Issues in Ergonomics Science*, 9(3), 223–254.

Morel, G., Amalberti, R., and Chauvin, C. (2008). Articulating the differences between safety and resilience: the decision-making process of professional sea-fishing skippers. *Human Factors*, 50, 1–16.

Nascimento, A. (2009). Produire la santé, produire la sécurité. Développer une culture de sécurité en radiothérapie. Unpublished doctoral dissertation, CNAM, Paris.

Nascimento, A., and Falzon, P. (2009, August). Safety management at the sharp end: goals conflict and risky trade-offs by radiographers in radiotherapy. Presented at 17th IEA Congress, Beijing, China.

Nascimento, A., and Falzon, P. (2012). Producing effective treatment, enhancing safety: medical physicists' strategies to ensure quality in radiotherapy. *Applied Ergonomics*, 43(4), 777–784.

Pariès, J. (2011). De l'obéissance à la résilience, le nouveau défi de la sécurité. Presented at Les entretiens du Risques 2011, Institut de Maîtrise des Risques, Gentilly.

Pariès, J., and Vignes, P. (2007). Sécurité, l'heure des choix. *La Recherche*, (Suppl. 413), 22–27.

Reason, J., Parker, D., and Lawton, R. (1998). Organizational controls and safety: the varieties of rule-related behaviour. *Journal of Occupational and Organizational Psychology*, 71, 289–304.

chapter eight

Courses of work and development

Corinne Gaudart and Élise Ledoux

Contents

How can ergonomics take part in designing courses of work that support the health of individuals and performance of companies? This question, in fact, leads us to define what is meant by the term *course*. Two main views can be distinguished. The first is related to employment, and likens the course of work to a career path and a profession, to the succession of positions or professional duties of a worker. The second is an individual and biographical view, where working hours are inserted, more broadly, into the course of life. The first view emphasizes the social and collective aspects of the course of work; the second emphasizes its individual and subjective aspects. The present chapter proposes a third view, whose goal is to join the first two together, based on an examination of courses of work from the point of view of activity, which is subjected to a process of ageing, viewed as an increase in the time experienced.

This view considers courses of work as being part of an individual and collective history – specifically, as occurring within multiple temporalities that combine the macro, meso and micro levels: temporality of public policy in matters of work and employment, which is more or less present depending on the country; temporality of management, which organizes the frameworks in which work is done; temporality of collectives that bear a specific view of their profession and its rules; and temporality of individuals themselves, who are engaged in the course of their own life. Therefore, advocating a constructive view of work based on 'courses of work' implies having a systemic view – a view integrating individual and collective aspects – following both a synchronic view focused on the 'here and now' and a diachronic view focused on evolutions. Here, 'understanding work to transform it' aims, precisely, to develop some reference points for the construction of courses of work. This involves carrying out analyses that place work activity at a crossroads between these multiple temporalities.

This temporal view leads us to analyze the activity deployed in work not only as a mobilization of the individual toward present action, but also as being part of the worker's course of life. When it is analyzed following a developmental perspective, activity reveals 'where the worker is standing', and the path that was travelled on. It also contains the future orientations of development. In this sense, activity possesses a historical dimension and is potentially open to the future. In ergonomics, this developmental view of activity takes its place in a body of research focusing on the relationship between ageing and work, where the course of work is seen as an increase in the time experienced (Molinié et al., 2012).

Several consequences can be derived from this definition. The individual temporality contains a biological temporality, directed toward decline, and a subjective, psychological one – the temporality of life events and of the interpretation of these events. These two dimensions do not live separate lives but intertwine together, setting themselves, through successive reconstructions, as resources or constraints for one another. Hence, this biographical course of life relates to a continuous process of transformation. The development of individuals takes place throughout the course of life, including during adult life, and therefore during working life. This process is multidirectional, in the sense that the direction of possible change is subject to great diversity.

Hence, the course of work reveals a constant reconstitution between processes of decline and growth. Processes of decline manifest themselves through the difficulties of the individual in adapting to his or her work environment and the changes that occur therein. They are the product of current working conditions, and of past conditions that may speed them up. The process of growth, in turn, refers to an increase in knowledge regarding the work environment. Following this view, the processes of decline and growth are closely interconnected. Experience is a product of this constant reconstitution. It is a process of construction that combines an accumulation of knowledge and know-how regarding the work environment, but also regarding one's own resources and constraints, i.e. how one can 'make use of oneself' in work.

This experience is set within the activity. It gives meaning to this activity, with regard to the course of work, and also guides it. Furthermore, the activity mobilized in work constructs this experience. Activity is mediated by skills (Teiger et al., 1998), which we will view here as means to regulate action between three temporalities: a managerial one, that is, sociotechnical in nature, and that refers to the goals and means of the company, as implemented through the organization of work; a collective one, which refers to work collectives, to peers and to the hierarchy, and that can therefore draw on several sources; and an individual temporality, as defined above, which is composed of biological, psychological and subjective dimensions. Therefore, courses of work are located at a crossroads between several individual and collective temporalities. And human activity, which takes its place within the course of work, relates to a process of regulation between these various micro, meso and macro temporalities (Gaudart, 2014).

Based on this, three major questions emerge: How can managerial and collective time frames pose themselves as resources or as constraints for these time frames? How can processes of decline and growth transform activity? Conversely, how can activity, in return, exert an influence on these multiple time frames? With this last question, we will arrive at the means for a constructive approach of work.

Managerial time frames yield a vision of the course of work as a process of decline

Whereas individual temporalities develop themselves within a combination, which is constantly renewed over time, between decline and experience, managerial time frames – those that design, organize and assess work – often have another vision of these time frames. This increase in the time experienced by workers is mostly associated with processes of decline rather than experience. This has direct consequences on the organization of courses of work, as workers draw close to the end of active life. This view of work as a decline is based on three different ingredients that combine with one another: social stereotypes that are present in society as a whole and are more or less fuelled by public policies at the national level, transformations of work that leave little space for the diversity of workers, and a view of skills as being based on prescribed work rather than on activity as it is done in the real world. This managerial view leads to mismatches between time frames (Alter, 2003). Managerial time frames and individual time frames do not develop following the same logic, and these conflicts may lead to declines in health over the course of life.

The first ingredient deals with negative social stereotypes related to ageing. One should point out that – in Western societies – these stereotypes remain quite stable over time. They even remain independent of the broader economic and social context. This has been tracked by several international inquiries carried out since the 1950s (Burnay, 2004). In spite of their dependability and professionalism, elder workers are suspected of being less able to adapt to change and perform well. These stereotypes are a feature shared between so-called industrialized countries (Rosen and Jerdee, 1997; Walker and Taylor, 1992). In countries such as France, the massive implementation of early retirement schemes has even led to moving forward the age at which one is viewed as being old, and therefore ill-suited to work. One becomes old not when one reaches the age set for departure, but a few years before (Guillemard, 1993). In concrete terms, this leads to a failure to invest in training for career development in the latter part of professional life.

This policy, which consisted in viewing the elder worker categories as a flexible variable – in particular to decrease the size of the workforce – has supported organizational and technological changes since as early as the 1990s. The quest for improved productivity, based notably on increasing flexibility, was pursued all the more easily when the population of workers involved in these transformations corresponded to the 'average worker' – neither too old nor too young, and in a reasonably good state of health (Gaudart, 2010). Problems occurred when the baby boom generation introduced a massive increase in the age structures of companies and the age of retirement increased, while the negative stereotypes related to age remained.

In addition to this social and demographic context, one must also consider the managerial practices that organize the management of human resources according to a skill-based logic, to increase flexibility in the workforce. This form of management of human resources and courses of work is accompanied by transformations of the work organization that are based on increasing the flexibility of production tools in order to increase productivity. In other words, increasing the flexibility of work tools implies increasing the flexibility in the workforce. Following this position, the workforce is viewed as a flexible variable that may be adjusted in terms of both size and skill. This logic of skill-based management is praiseworthy from the point of view of its initial project for development. However, when it is subjected to a strong economic rationale, it tends to confuse acquired skills with required skills, and to follow a short-term view. Furthermore, it tends to reduce skills to technical knowledge and know-how, setting aside the tricks of the trade and the processes of individual and collective regulation that make it possible to 'get the job done', in terms of both quantity and quality (Pueyo, 2012). This twofold confusion is not without consequence. On the one hand, the issue of developing experience over time remains invisible. On the other hand, this logic of required skills tends to 'reset the system' after every organizational or technological change. Considering this situation at the individual level, and also considering the fact that the ageing process leads to an increased need to combine decline and experience, this twofold confusion may lead to adverse impacts on health.

Let us take an example that illustrates these temporal mismatches between the managerial temporality and the individual one, and their consequences. This example is derived from some research work that reviewed 20 years of organizational changes within a company in the automotive sector (Gaudart, 2000a; Gaudart and Chassaing, 2012). During the 1990s, this company developed two types of management tools in order to improve its productivity: making each worker accountable for the quality of work, and developing job rotation as a dominant model for the course of work. One of the main difficulties encountered by this company lies in the fact that its age structure is gradually ageing – more specifically, it tends to aggregate toward intermediary ages – and that workers encounter difficulties with respect to versatility as early as 40 years of age. Indeed, beyond this age, more than 50 per cent of workers occupy only one position in the company. In other words, difficulties in coping with working conditions appear as early as 40 years of age, and these difficulties are first diagnosed by the management as being related to age: problems in adaptation and learning. Analyzing the activity of these 'elderly' operators reveals that the reality of the situation is more complex. They develop operative strategies that allow them to preserve themselves from time-related and physical constraints. This experience does, indeed, take into account, in a combined fashion, the processes of decline and experience in the professional environment. However, these strategies remain dependent on workstation design and on the variability of the prescribed setup. Therefore, versatility may call into question the construction of such strategies. Increasing the accountability of workers with respect to quality leads to a fear of error, and to elder operators making trade-offs in favour of this criterion, rather than in favour of strategies for self-preservation. Professional prospects, for both young and elder workers, are set when leaving the production line: to the top for younger workers, who aim, for example, for a management-level job; and to the bottom for elder workers, through the recognition of an occupational incapacity.

Ten years later, this same company has gone on to set up a new management policy that emphasizes the importance of quality in productivity through the standardization of gestures and the implementation of tools to ensure that they are adhered to. The room for manoeuvre of elder workers is further reduced. The extension of productivity (kaizen) workshops then caps off an already well-established policy of rationalization. The production lines have improved in performance, but the positions have become increasingly selective. An increasing proportion of the population of production workers is confronted with medical incapacities. This further complicates the organization of 'courses of work'. These courses are organized by default, i.e. based on what positions are compatible with the medical incapacities.

This example illustrates some typical mismatches between the managerial and individual temporalities. One can clearly see here that stages of change are critical times where the combination of these temporalities is played out, over and over again. The results of this process set the foundations for courses of work that are yet to come. When they are not overcome, mismatches between temporalities lead to declines in health that obliterate the course of development. They emphasize the importance of the processes of decline, and may even accelerate them.

The temporalities of collectives play the part of an interface

Work collectives are the bearers of a history, rules and values of the trade, which are transmitted from generation to generation – and more broadly from elder to younger workers. In that sense, these collectives have a normative power, which regulates work toward immediate action, but also in terms of the course of work within the collective. Therefore, their position is that of an interface or 'double address' (Clot, 2009): they may resist in the face of managerial time frames, whose views are dissonant with the values and rules supported by the collective; they are also positioned toward the individual time frames that make them up, and may play the part of resources or constraints with respect to these individual time frames. From the point of view of age (Cau-Bareille, 2012), they might thus define the rules for integrating new workers in the processes of reception and learning, and transform these rules based on the diversity of individual time frames that compose them. Several works in ergonomics research (Marquié et al., 1998) thus suggest that in collectives composed of workers of various ages, bartering may occur to exchange arduous work conditions with experience of the trade. Thus, the collective time frames occupy a position that was deserted by managerial time frames, when these time frames adopted a view of ageing as an increase in 'time lived through'.

As regards this role of double address, managerial temporalities and individual temporalities put the normative power of collective temporalities to the test, with respect to their ability for resistance and transformation. Caroly (2012), when carrying out a research project on the emergence of occupational incapacities of post office workers, showed how work collectives may pass or fail this test. Collective work is a key element in a work collective that supports the individuals that form it. We will continue here to view the problem through the lens of the relationship between age and work to examine in greater detail the various roles played by collective temporalities in their position as an interface.

The lack of a collective

Collective temporalities may find themselves unbalanced by managerial temporalities. The lack of a collective highlights its potential role as a buffer. When this collective is absent or disrupted, managerial temporalities are in direct contact with individual temporalities. At this point, one can find adverse effects on health, which have already been discussed earlier (Caroly and Barcellini, this volume). Caroly (2012) analyzed how, in one of the post offices studied in her work, the lack of a work collective and of any collective work led to an increase in turnover and restrictions of aptitude.

Organizational and technological changes thus remain stages of increased fragility for individuals and collectives. In a public institution in charge of distributing family credit and social benefits, a change in the software used to process files led to conflicts of values in technicians, specifically regarding what constitutes 'quality work' (Gaudart, 2000b). The management anticipates possible difficulties of learning in these ageing technicians, whose professional experience was built based on a different tool. Thus, it reduces their skills to technical skills only, believing that their problems can be solved with adequate training. However, the old software program plays a much more important part than being a simple technical tool. Its design mediates the collective rules of the trade, which are based upon an idea of what constitutes public service. The new piece of software, which was designed following a managerial rationale of counting technical actions, stops this quality work from being performed in the activity of workers. The problem is not so much one of training than it is one of design. Faced with a rationalizing design that individualizes work, the collective becomes disrupted. Many technicians tend to stand back and remain in the background by yielding to prescribed work. Several months after the implementation of this new software program, new forms of collective work reappeared. However, this ordeal also left many workers 'on the side of the road' and estranged from the collective.

An organization that 'manages human resources'

Collective temporalities may also exhibit a strong ability for resistance, going so far as to stand in for the functions of human resources. In the 1990s, the sector of steel manufacturing in France underwent a radical change in the organization of its courses of work. It is one of the earliest sectors to have implemented a skill-based logic as a part of joint agreements. These agreements aimed to manage the major transformations that the sector had gone through, in technological, economic and demographic terms. This was achieved by redirecting training programs toward its specific needs, and by offering workers beneficial career development schemes. The issue of age is central in these agreements. The fact that the sector has had to cope with a considerable increase in the duration of professional life implies that these agreements should offer means for the development of skills right up to the age of retirement, and not resort to sidelining workers.

The ways in which these agreements should be implemented is left at the discretion of the production sites. In one of the sites involved (Gaudart, 2003), the steel manufacturing unit decided to merge two departments, one involving the operators of travelling cranes, and the other involving workers in charge of ladle maintenance. The idea was to propose a development of job rotation between these two trades, as a solution for developing career paths. This did not work. Because the operators were all elderly – the youngest was 40 years old – this result was attributed by top-level management to a lack of motivation and to problems in learning on the part of the workers. Activity analysis revealed, on the one hand, differences in the collective temporalities between these two trades, and on the other hand, the mismatches of these two temporalities with respect to the managerial time frame.

Let us first examine the case of travelling crane operators. The time frame defines rules for progressing in the course of work, following two scenarios, both of which foster single-purpose work. The first scenario is organized based on a view of age as a process of decline. Ageing in crane operators is accompanied by an increase in osteoarticular disorders. In a context of high temporal constraints, notably because of the managerial wish to increase flexibility in the production system, some operators, after having occupied many different positions, settle on one crane that suits them best, i.e. a crane that they can truly control in the face of production constraints, and that allows them to protect themselves. The second scenario highlights the function of experience increasing with age. Single-purpose work tends to focus on the stations that are said to be the most difficult to hold, those combining a diversity of tasks with strong time constraints. The operators who work there are those whose expertise in the use of a crane allows them to ensure reliability. In fact, this expertise implies transgressions of safety rules, supported by the work collective and by middle management. Because of this, the expertise successfully resists the requirements of the managerial time frame. Furthermore, the organization based on single-purpose work is related to a history of the trade over several generations, dating back to the times where new operators rotated, i.e. were versatile, until they 'found their place', i.e. their own crane. This single-purposeness was, then, viewed as a sign of expertise by the collective.

The preservation of each worker

A homogeneous composition of the members of a collective, in terms of both age and course of work – we will call these workers elderly – may foster the development of a collective emphasizing a function of preservation. This function takes its full importance when past and current working conditions are particularly tough. The second collective temporality, concerning ladlers, has these features. In the face of physically strenuous tasks – both those of the past, which have had effects on their health, and those of the present, which involve intense physical effort in a difficult thermal environment – ladlers share the work between themselves following two regulatory processes. Two teams of two operators are formed, each team carrying out in turn the maintenance of a ladle while the other team rests. Within a single team, the shared experience allows workers to perform well-managed cooperative work, reducing the physical strenuousness of work. As we have pointed out, this type of organization is viable only on the condition that there exists a collective that is both stable and homogeneous. This collective is put to the test by the development of versatility between trades, in the sense that this new organization of work has led to the introduction of 'new' workers, who are not familiar with this shared history. Is this kind of collective, which is geared toward taking into account the processes of decline, truly desirable? In a short-term vision, it demonstrates a great degree of usefulness, since it makes it possible to avoid more serious damage to health. However, this stable, homogeneous collective, which is a product of earlier policies of age management, can also be seen as a collective that is 'locked onto itself' and whose function is not to promote health, but to preserve it.

In the examples we have quoted above, the managerial temporalities do not play against the individual time frames. However, they still carry a heritage that is loaded with more macro time frames, with models for the management of work and human resources – in ergonomics, we talk about a model of man at work – that are poorly sensitive to the determinants and to the meaning of activity. Age, in the sense that the issues of experience and decline take centre stage, highlights the mismatches that situations like this can lead to. Because of this, age can also turn out to reveal the means of action that should be implemented. Here, collective time frames play the essential role of acting as an interface between the managerial and individual time frames. They also play an important part in the transformative dynamics that are a focus of our interest.

Supporting and accompanying the design of developmental situations

Let us return to the initial question: how can ergonomics take part in the design of courses of work that support both the health of individuals and the performance of companies? Following a constructive approach to work, the actions that ergonomics can propose in terms of courses of work cannot take the form of specifications that might describe the ideal course.

The goal is, instead, to provide resources to the various temporalities involved in order to alleviate mismatches. The developmental approach of individual temporalities that we advocate here is an approach that ensures continuity in work between the temporal categories of the past, present and future. In other words, following a constructive ergonomics approach, an activity that makes it possible to develop oneself is the activity that allows, in the present time of activity, drawing experience from the past while opening up a space of possible futures (Gaudart, 2014). The question then becomes: what are the tools that are capable of supporting this developmental view?

Providing memory to management

Managerial temporalities are also collective trade-based time frames that are subject to the consequences of policies to increase the flexibility of work. Such policies promote a short-term vision that has led to memory loss. The increased mobility of top management is undoubtedly one of the chief causes of this phenomenon, with an injunction for immediate performance. How can the managerial time frame support a developmental view of courses of work, when it is itself stuck in the present time?

Ergonomics can support this 'work of remembrance'. In the long term, the practitioner in ergonomics takes on a job that is close to that of a historian, in order to put back in context the major managerial actions of the company. In the short term, another essential role of ergonomics is to contribute to increasing the visibility of the managerial mindset that led to the emergence of the problem for which the services of the ergonomist are required. The pressure, introduced by a managerial mindset, for immediate returns on investment, often leads to 'losing the story' – or at the very least, losing track of the meaning behind actions. In a petrochemical company that exploited gas reserves (Gaudart et al., 2012), a reduction in resources naturally led to reducing the workforce by selecting not to replace workers following their retirement. This decision was not related to financial constraints. It seemed to be a logical decision, which had its own rationale. However, the difficulties that this decision led to later on, regarding how everyday work was carried out, and regarding the risks incurred, led management to call upon the services of ergonomists. Furthermore, the managerial and collective temporalities clashed strongly with one another on this occasion, leading to a walkout. Reconstructing the decision processes that had been involved turned out to foster some crucial dynamics. This reconstruction took place, following a report by the ergonomists of previous managerial decisions, when the director turned to his team and asked: 'By the way, why did we decide to cut down the workforce in the first place?'

The reconstruction of managerial memory can also take place through the construction of tools that make it possible, on the one hand, to analyze the work as it is undergoing evolutions, and on the other hand, to connect the knowledge of human resources with the knowledge of production. Reflections about courses of work – as we have defined them above – cannot take place without connecting these pieces of knowledge together. Yet, they are dispersed within various functions of the company – human resources, occupational medicine and production. Work demographics (Molinié and Volkoff, 1998), in the sense that they combine data related to age, seniority and features of work-related evolution, are a fundamental tool to plan the relationships between work and health, and to understand the selective effects of work. They are also a tool to plan the future, since they allow the use of projections as intermediaries.

Transmission as a process aiming to regulate mismatches between temporalities

The point above aims to ensure that the past becomes present, and that the managerial temporality is given a form of situated memory, to open up a debate about the meaning of actions carried out in the past. This implies that a developmental approach to courses of work, whose goal would be to support past experience in the present time of activity, should remain a key stage to plan the future of those courses.

Transmission – which we will define here as a process of mutual sharing of trade-related knowledge and know-how between new and elder workers – is a second potential means to open up courses of work to the future. Development also means being able to turn constraints into resources. In this sense, the baby boom generation, which is often presented as a negative influence on organizational performance, might instead be viewed as an opportunity for the design of courses of work. Ageing in this generation, as it is currently happening, has made it possible to consider the diversity of ages at work that it has produced, to plan courses of work in a different way, that is, to replace retiring employees with newer employees, whether these new employees are young or not, but also to deal with the issue of the extended duration of professional life. Therefore, this diversity in the age structure can become a resource, provided there is some form of support on the part of the managerial temporality to ensure a link between past, present and future.

Following this view, the work collective plays a fundamental part in this process. In a gerontology ward in a hospital (Gaudart and Thébault, 2012), the arrival of a new nursing auxiliary in training led to a dynamic exchange of views regarding the rules of the trade – here, specifically, about how to deal with an elderly patient. This went far beyond the integration of the newcomer, which typically occurs in the form of two-person teams that include a new and a senior worker. The work collective, that had been renewed following the departure of several senior nurses and major organizational changes, took hold of this training opportunity to 'come to an agreement'. The integration of this novice was no longer the business of a single two-person team, but of the collective as a whole: the collective loosened up its time constraints to provide some additional quality training time, and provided its members with the possibility of joining the pair of workers. The involvement of a nursing auxiliary, newly arrived in the department, whose own work history has exposed her to an extensive experience of caring for elder patients, sparked off a debate regarding the connection between the technical aspect of work – *cure* – and the psychological aspect – *care*. From then on, the situation of welcoming a student became a special time for discussing the past experiences of each auxiliary, the conditions of transmission, and to design the rules of the trade to plan out courses of work for every auxiliary in the ward. In a sense, transmission became a developmental task.

This role of the collective becomes all the more crucial when employment is affected by low job security, as is the case in the cinema industry, which primarily employs freelancers. A study carried out in a population of cinema technicians showed that the lack of job security affected the transmission of knowledge in various ways (Cloutier et al., 2012). Considering the fact that most knowledge is acquired on the job, the requirements of the industry concerning novice workers are paradoxical. The expert workers who were met agreed together that progressing in the industry is a long process, comprising many stages, that trade knowledge is acquired through a confrontation with various situations and through repetition, and that the process requires a lot of time. However, the way in which the cinema industry is organized, including the lack of job security and fierce competition that characterize the trade, forces novices to 'prove their worth' very quickly. A novice's 'performance' is often assessed based on only one day or even only a few hours' work. Indeed, although the educational, social and professional history of the worker does play a part in the integration, the strongest factor of integration is subscribing to the culture of the trade. This relates strongly to respect for the hierarchy and for the rules of the trade, such as mutual aid and adaptation to modes of communication. The more the novice worker adopts behaviour that is true to the culture of the trade, the faster he or she is taken in by the work collective. It is access to this collective that will then open up opportunities for learning through transmission. The transmission of this 'culture of the trade' becomes a key, both to accessing means to learn the trade and to being hired in future productions. However, this lack of job security can also become a major obstacle for transmission on other levels, since the various protagonists involved tend to have to compete against one another when looking for employment in other productions.

Placing the issue of the links between past, present and future at the heart of a developmental approach of courses of work leads us to revisit the functions of activity and activity analysis. Activity contains, in itself, a potential for development, in the sense that it takes place within a field of experience and a horizon of expectations (Koselleck, 1990). Activity, in the present time, makes use of the past and opens up a space of possible futures. One of the functions of activity, then, lies in its ability to create time – a time for oneself (Sivadon and Fernandez-Zoïla, 1983). In the course of work, this individual temporality connects and confronts itself with other ones, bearing other different rules for the production of time – the rules of management, the rules of the collective and other, more macrosocial rules. This connection materializes in the activity, defined here as a process of regulation between these multiple temporalities. Following this view, a constructive approach to work, such as can be borne by ergonomics, would be an approach seeking to transform unfavourable situations – where these temporalities do not connect well with one another – in sustainable alliances, turning a dyschronous environment into an enabling environment. This environment then has two interrelated functions: to allow the activity to circulate between past, present and future, and to support the creation of a sustainable alliance that is constantly transformed over time from the influence of the changes that occur in each of the temporalities. This constructive ergonomics, then, is one that views activity analysis following a systemic view, carried out in a diachronous and synchronous fashion, and whose scope of action lies in connecting these various temporalities together.

References

Alter, N. (2003). Mouvement et dyschronies dans les organisations. *L'Année Sociologique*, 43(2), 489–514.

Burnay, N. (2004). Les stéréotypes sociaux à l'égard des travailleurs âgés. *Gérontologie et Société*, 111, 157–170.

Caroly, S. (2012). Gestion collective de situations critiques au guichet en fonction de l'âge, de l'expérience et de l'organisation du travail. In A. F. Molinié, C. Gaudart, and V. Pueyo (Eds.), *La vie professionnelle: age, expérience et santé à l'épreuve des conditions de travail* (pp. 223–234). Toulouse: Octarès.

Cau-Bareille, D. (2012). Travail collectif et collectif de travail au fil de l'âge: des ressources et des contraintes. In A. F. Molinié, C. Gaudart, and V. Pueyo (Eds.), *La vie professionnelle: age, expérience et santé à l'épreuve des conditions de travail* (pp. 181–203). Toulouse: Octarès.

Clot, Y. (2009). Clinic of activity: the dialogue as instrument. In A. Sannino, H. Daniels, and K. D. Gutiérrez (Eds.), *Learning and expanding with activity theory* (pp. 286–302). Cambridge: Cambridge University Press.

Cloutier, E., Ledoux, E., and Fournier, P. S. (2012). Knowledge transmission in light of recent transformations in the workplace. *Relations Industrielles*, 67(2), 304–324.

Gaudart, C. (2000a). Conditions for maintaining ageing operators at work – a case study conducted at an automobile manufacturing plant. *Applied Ergonomics*, 31(5), 453–462.

Gaudart, C. (2000b). Quand l'écran masque l'expérience: changement de logiciel et activité de travail dans un organisme de services, *PISTES*, 2(2). Retrieved from http://www.pistes.uqam.ca/v2n2/articles/v2n2a4.htm.

Gaudart, C. (2003). La baisse de la polyvalence avec l'âge: question de vieillissement, d'expérience, de génération? *PISTES*, 5(2). Retrieved from www.unites.uqam.ca/pistes.

Gaudart, C. (2010). Les âges au travail. In L. Théry (Ed.), *Le travail intenable – résister collectivement à l'intensification du travail* (pp. 133–149). Paris: La Découverte.

Gaudart, C. (2014). Les relations entre l'âge et le travail comme problème temporel. *PISTES*. Retrieved from http://pistes.revues.org/2829?lang=en.

Gaudart, C., and Chassaing, K. (2012). Formation *"in situ"* et "école de dextérité" dans l'automobile: analyse des modalités d'apprentissage et de leurs coûts pour les opérateurs. In A. F. Molinié, C. Gaudart, and V. Pueyo (Eds.), *La vie professionnelle: age, expérience et santé à l'épreuve des conditions de travail* (pp. 75–94). Toulouse: Octarès.

Gaudart, C., Garrigou, A., and Chassaing, K. (2012). Analysis or organizational conditions for risk management: the case study of a petrochemical site. *Work*, 41, 2661–2667.

Gaudart, C., and Thébault, J. (2012). La place du care dans la transmission des savoirs professionnels entre anciens et nouveaux à l'hôpital. *Industrial Relations*, 67(2), 242–262.

Guillemard, A. M. (1993). Travailleurs vieillissants et marché du travail en Europe. *Travail et Emploi*, 57, 60–79.

Koselleck, R. (1990). *Le futur passé. Contribution à la sémantique des temps historiques*. Paris: EHESS.

Marquié, J. C., Paumès, D., and Volkoff, S. (Eds.). (1998). *Working with age*. London: Taylor and Francis.

Molinié, A.-F., Gaudart, C., and Pueyo, V. (Eds.). (2012). *La vie professionnelle: age, expérience et santé à l'épreuve des conditions de travail*. Toulouse: Octarès.

Molinié, A.-F., and Volkoff, S. (1998). Elements for a demography of work. In J.-C. Marquié, D. Cau-Bareille, and S. Volkoff (Eds.), *Working with age* (pp. 71–90). London: Taylor and Francis.

Pueyo, V. (2012). Quand la gestion des risques est en péril chez les fondeurs. In A. F. Molinié, C. Gaudart, and V. Pueyo (Eds.), *La vie professionnelle: age, expérience et santé à l'épreuve des conditions de travail* (pp. 257–284). Toulouse: Octarès.

Rosen, B., and Jerdee, T. H. (1977). Too old or not too old. *Harvard Business Review*, 7, 97–108.

Sivadon, P., and Fernandez-Zoïla, A. (1983). *Temps de travail, temps de vivre*. Brussels: Pierre Mardaga.

Teiger, C., Cloutier, E., David, H., and Prévost, J. (1998, September). Le temps de la restitution collective des résultats de recherche dans les dynamiques de l'intervention. Le cas du travail de soins à domicile au Québec. Presented at 33rd SELF Conference, Paris.

Walker, A., and Taylor, P. (1992, May). The employment of older people: employers attitudes and practices. Presented at ESRC Conference Past, Current and Future Initiatives on Ageing, London.

section two

---

*Dynamics of action and
dynamics of development*

chapter nine

Interventions as dynamic processes for the joint development of agents and organizations

Johann Petit and Fabien Coutarel

Contents

The environment and the individual do not constitute separate entities, isolated from – and forced to coexist with – one another. Instead, they are two elements that evolve jointly in a search for balance. These general considerations about the human condition apply to the world of work. The elements of the work environment are major determinants of the activity of the individuals involved. They are also targets, in the sense that the environment is an object of transformation for the individual. Within these determinants, organizations have, over the past few years, been the subject of much attention on the part of ergonomists, since it remains a source of influence and a major target for the activity. Because of this, organizations have become an object of study and transformation for ergonomists. To specify, this text will argue that ergonomic action can be viewed, on the one hand, as a transformation of the organization, and on the other hand, as an opportunity to influence the development of individuals.

In more concrete terms, in this chapter we will describe the course of an ergonomic intervention, its results and the reflections it led to. Over the course of this intervention, which was centred around an organizational change, we had the opportunity to set up an 'experimentation' that allowed us to solve the actual problems of work and to construct the forms of collaboration that were necessary to solving these problems. This allowed us to identify some essential features in order to structure a system for the ergonomic intervention. We organized this system around moments where it was possible to define the rules of operation of work, and moments where more space was left for the development of individual and collective experience. This last point allows us to liken our method to an 'intrinsic approach' (Rabardel and Béguin, 2005) – an approach that is consistent with the personal points of view of the persons that make up the situation, that is, the workers themselves. This element seems crucial to us, since it allows individuals undergoing a process of organizational transition that is synonymous with cognitive and social disruptions to construct their own spaces for learning. By working on finding solutions to everyday difficulties related to their work, and attempting to construct a system that is as effective as possible for dealing with these problems, workers construct new representations of their current and future work. This allows us to open up two paths for reflection:

- On the one hand, the running of the organization must be part of a developmental process, by allowing a joint evolution of the organizational structure and of human activities.
- On the other hand, an ergonomic intervention in organizational design may take place as part of this process.

The organization of development?

The production of quality work is a major issue for any working individual. However, in order to produce quality work, workers must be capable of action (Rabardel and Pastré, 2005) – that is, of adapting their modes of operation to every situation. This requires that the workers' environment, especially the *organizational* environment, their skills, and their state – from the physical, cognitive and psychical points of view – allow them to do so. When the operator is forced to only apply the strict prescription, whereas the situation suggests the need to innovate in order to deal with particularities, the consequences are twofold. On the one hand, the work may not achieve its intended results, and on the other hand, the worker may be prevented from exploring new skills, which will restrict his or her cognitive development, leading to a potential deterioration of his or her psychical state (Clot, 2008; Davezies, 2008).

In order to develop, the skills of the workers must also be open for discussion and confrontation, since the experience lived by each individual takes on its full value if it acquires a collective dimension; the sharing of experience opens up prospects for the development of skills in other individuals. To achieve this, it is crucial that the complexity of work should not be denied at all levels in the organization, in the form of organizational silence (Morrison and Milliken, 2000). On the contrary, this complexity should be open to discussion, and also for use in order to adapt the organization. In other words, the concrete difficulties encountered by the workers should potentially be an object of debate with colleagues and with the hierarchy. By allowing workers to entertain debates about their work, one can foster the empowerment that is essential to the development of individuals and of the organization.

In order to address the organization, the theory of organizational regulation, derived from work carried out in organizational sociology (Reynaud, 2003; de Terssac, 1998), constitutes an appropriate starting point for our argument. This sociological approach to organizations leads us to view the organization as a living system with two faces: on the one hand, the organizational structure, which is composed of procedures, tasks, goals and organization charts, and on the other hand, human activities. From there, running the organization involves maintaining a balance, allowing human activities and the organizational structure to evolve jointly. This approach to organizations proves to be essential to the work of ergonomists. It allows an understanding of the level of negotiation that is required between the activity of workers and the organizational structure in order to achieve work goals. Furthermore, from a point of view of health and efficiency of production, this approach to the running of an organization allows us to define the operational leeway that is necessary to operators (Coutarel, 2004; Hasle and Jensen, 2006; Vézina, 2010). Therefore, designing a reliable organizational system requires taking into account the fact that the job of the agents of the system consists of producing the organization. Following this view, one cannot consider organizational change as the adaptation of individuals and collectives to a new structure (Caroly, 2010). Organizational change relates to a joint adaptation of the structure and of human activities, both individual and collective (Béguin, this volume). This joint adaptation in favour of organizational change can only be achieved – and can only be efficient – if it allows a development of the activity of individuals and collectives, through the acquisition of new knowledge and skills in the future structure. In other words, organizational change must be viewed by the workers as a resource that can allow them to take part in the construction of local responses to their specific problems.

The challenge for ergonomic action is here. How can one give structure to plasticity in the organization in order to take charge of contingencies, and to allow operators to take part in the constant redefinition of the organizational structure in a developmental perspective?

The substance for the remainder of the discussion will come from data derived from an ergonomic intervention. Although the results discussed in this chapter are based on about 10 different interventions, we will focus on only one of those for the purpose of clarity.

Organizational change in a health insurance firm

The context

The intervention took place in a health insurance firm, numbering approximately 3000 employees deployed all over the French territory, within about a hundred departmental branches. The main activity of this firm consists of dealing with the management and reimbursement of healthcare fees. The demand for the ergonomic intervention came from the head office, which required support for an organizational project that was already underway. The main goal of the project was to 'improve the quality of service', by rationalizing production through a strict separation between back office and front office activities. To achieve this, processing centres had been created to manage the massive numbers of files to be processed. When the demand was formulated, the processing centres were already implemented, as illustrated in Figure 9.1. However, the overall running of production did not satisfy management (low productivity and loss of quality).

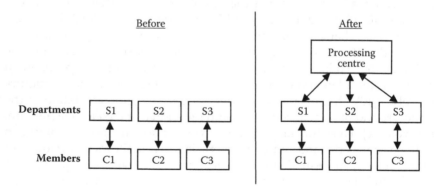

Figure 9.1 Overall organization, before and after the change.

Analysis of work

In a first stage, we carried out analyses of work in two departmental branches. We quickly realized how much the implementation of the processing centres had had negative consequences on the running of individual branches. Indeed, files that could not be processed in the processing centres were returned, and monopolized a considerable amount of work time of the departmental agents. Because of this, the work of the ergonomist focused more specifically on the processing of these rejected, and often more complex, files. The difficulties related to the settling of health expense sheets had been underestimated. Indeed, no two claim sheets are exactly alike. They are subject to some variability, which cannot be easily quantified. Yet, it seemed to us that the hypothesis 'settling a file means inputting the claims forms and authorizing payment' was at the source of part of the organizational choices. In fact, a large number of files were 'rejected' by the processing centre and sent back to the departmental branches because the former were locked in an 'organizational impossibility' regarding the management of this variability.

Furthermore, 'settling' often means following a 'story', or even a 'saga'. Yet, in the organizational configuration in which we carried out these analyses, 'following a story' seemed to be a more difficult thing to implement. Indeed, although the earlier organization allowed the group to follow those of its subscribers who encountered the most problems, following this history proved to be much more difficult after the implementation of the processing centres (different contacts with the same subscriber, difficulties for each subscriber to maintain contact with a single contact in the group): one often has to 'start again from scratch'. Workers had little access to computer support for tracing the history of a subscriber, and paper records were the only true witness of the whole story: there was less information in the computers than in archive boxes. These are the reasons why the processing centres caused a loss of traceability regarding subscriber files. In fact, the people who had the information were not the people who were called upon when one wanted to use it. Finally, in this context, standard files were processed more quickly by the processing centres, but more complex files (concerning subscribers with more difficulties) were processed in a longer time frame. As a consequence, according to the workers and branch chiefs, the processing centres did not provide the expected assistance in terms of 'production support'. The consequences of this were an increase in the error rate, the processing times and the number of complaints regarding quality of service. The complaints and absences of employees, on the other hand, were rooted in part in the gradual degradation of the quality of their work. In fact, everything happened as if aiming to improve the quality of service offered to subscribers following criteria of productivity had caused new forms of degradation of that quality. The organizational structure seemed, in a number of cases, to go against the possibilities for regulation to cope with malfunctions, altering the quality of service provided to the subscriber. This organization seemed to have 'forgotten the subscriber'.

This opened up a reflection about the contents of our intervention, with and for national and local managers. From there, we agreed that improving the performance of the processing centres implied allowing them to process the variability of claims forms and requests, by restoring interactions between them and the departmental branches, in order to place the concept of the history of a subscriber and file at the heart of the activity. At this point, we also highlighted the fact that front office and back office workers had unofficially compensated for some weaknesses of the organizational structure with local adjustments. Yet, what seemed crucial to us was that these initiatives were extremely precarious, in the sense that they led to very informal agreements between structures – agreements that were then subject to power games (departmental branches gave orders to the processing centre, which was 'billed' as a service provider). In this context, the organizational prospects regarding both the processing centres and the departmental branches needed to be decided based on choices that had not yet been explored. This exploration could rely on improving the autonomy of the processing centre and increasing the contacts between the department branches and the processing centre, in order to facilitate the processing of files.

Elaboration of a system

This is why we elected to set up a local work group in order to solve the malfunctions. Based on exchanges with a wide range of stakeholders (agents and managers from the departmental branches and from the centre), we quickly arrived at the idea of simulating specific forms of organization in order to optimize the processing modes of these files. Finally, after having addressed specific choices, we submitted at the national level the possibility of experimenting with these choices in real time. This allowed us to solve very concrete malfunctions by suggesting solutions that were simple to implement. We first worked with 3 departmental branches and 1 processing centre, then gradually with about 20 departmental branches. This work lasted for 18 months and allowed us to experiment in real time with solutions to the various malfunctions highlighted in the diagnosis performed by the ergonomists as well as other malfunctions that appeared over time. Some solutions took the form of specific procedures between a branch and the processing centre, while others were more related to procedures or more general organizational structures, influencing each of the 20 branches. This entire stage of simulation-based solution-finding, coupled with real test stages, quickly took on both the name and the form of an experimentation on the organization of production between the back office and the front office. We were able to pursue this experiment for an extended period of time, notably because it yielded tangible results.

The files that had been rejected by the processing centre very quickly became an object of work. In quantitative terms, they represented about 200 files per day and per branch. In qualitative terms, these were either commonplace files that had been returned with the wording 'incomplete' or 'missing a signature', or more complicated files for subscribers who were encountering financial or social difficulties. The experiment allowed us to concentrate on these 'problem files' in order to decrease their numbers and to allow the emergence of more general topics, such as the lack of knowledge of other agents' work, the lack of articulation between modes of execution for a single, same file, the sharing of skills, the activity of settling a file, the quality of service, etc. This collective work regarding the 'remodeling' of existing rules led to a decrease in the number of rejections to 25 per day, also allowing us to improve part of the service provided (quicker processing times) and the ability of the processing centre to help the local branches (skill development).

The idea of reducing the rate of rejections had been clearly identified, but the means needed to achieve this were far less clear. Over the course of several meetings, the topics of debate broadened, trust was established between the stakeholders who were present, and one could imagine a more global vision for the processing of files between entities, and by way of consequence, open up for debate a more wide-ranging view of the service and quality of service. There was no longer any doubt that to reduce the rejection rate, the processing centre should gain some time to deal with the files.

Results of the intervention

Overall, these results also led to a decrease in the number of complaints by subscribers. Freeing up this time also allowed the processing centre to increase its capacity for dealing with files. It therefore became able to deal with more complex files, which were not previously processed. Its scope of action broadened over the course of the experimentation. This change was perceived as a form of recognition by the processing centre staff, through the diversification of tasks and, above all, through a more consistent construction of the trade regarding the issue of settling files, and finally, as a solution to the development of new tasks into departments.

Whereas the processing centre had previously been viewed as an agent that disrupted work, by the end of the experiment, it was viewed rather as a structure with which it had become possible to construct a new identity for one's work. Indeed, this entire work focusing on file rejections, which had seemed very basic and technical, allowed us to apprehend and process complex issues related to change, identity at work, the development of skills or the recognition of the work of other agents. At the national level, the processing centres were subsequently recognized as service centres, on par with the different departments. This change, which was more symbolic than operational, suggested the appearance of a feeling of federalization of the processing centre, which was considered up to then as an external service provider. As a background to this evolution, the debates regarding work situations allowed exchanges regarding the various practices of the people involved. The various work groups functioned relatively independently from the national level of management, which was called upon to make decisions that related to its jurisdiction. In order for our action to be effective, a certain degree of reactivity in decision-making was required. However, the usual channels (hierarchy, technical and administrative services) did not offer the expected reactivity. Our position allowed us to construct an informal network of decision-making, allowing this reactivity. However, the work carried out in these groups also pointed out a weakness in the decision-making process, specifically between the national and local levels, which group participants expressed regularly. During the closing meeting of the experimentation, beyond a general acceptance, by the stakeholders involved, of the results obtained in the group sessions, a revision of the regional workings of the departmental branches and the processing centre was considered. The regulation process that was implemented had become necessary to all of the workers, in order to cope with the malfunctions encountered between the departmental branches and the processing centre. Indeed, each branch – along with the processing centre – could then find working responses to everyday problems. Therefore, a region-based organizational structure was proposed, which was defined by rules of operation, key stakeholders and a manager.

Discussion

Designing an operative system

An essential aspect of this approach focused on the modes of cooperation that made it possible to carry out the transformations and their experimental implementation. The evolution of modes of cooperation is assessed based on proposals to transform the organizational structure. In more concrete terms, the goal was to structure the organization in such a way as to allow workers and managers to retain an ability to act upon the rules – in de Terssac's (1998) words, to structure the 'work of organizing'. The structuring proposed by the participants emphasized two significant elements in the evolution of modes of cooperation between the departmental branches and the processing centre (Figure 9.2):

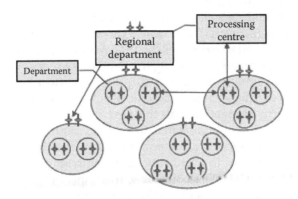

Figure 9.2 Final proposal for the regional organization.

- The central role ascribed to the processing centre in managing regulations at a regional level
- The establishment of a hierarchy of cooperation (between two branches, and between a branch and the processing centre)

This structuring led to various modes of regulation, depending on the case: 'hot regulations' (de Terssac and Lompré, 1996), which were not formalized, between branches or between a branch and the processing centre, and more formal 'cold regulations', related to the officialization and homogenization of rules for the whole region. Furthermore, in this way, the regional head office recorded the features of the groups that had been set up for the experimentation, which allowed a form of continuity in the types of exchanges that had built up over the course of the experimentation.

However, some other elements of cooperation were also proposed. These involved the connections between regional and national management. Indeed, the various decisions that were made at the regional level translated to a local decision, which had to be validated by persons in charge at the national level. This organizational structuration altered quite noticeably the forms of cooperation between the front office and the back office involved in the production of a service. It later served, in the overall process of change management, as a reference to define a blueprint for the benefits of society at the regional level. The experimental setup allowed solving the difficulties related to work, which was an essential point for agents of departmental branches and for agents of the processing centre. However, it also made it possible to reflect upon a new form of organization at the regional level, between the departmental branches and the processing centre. Leaders at the local and national levels had identified this possibility quite well. Furthermore, the organizational structure that was suggested in the end seemed to exhibit some features of an operant model (Wisner, 1972), in the sense that it was a model of the situation that represented some essential aspects of the real world, that allowed objective measures and was capable of leading to effective solutions. In this respect, the experimental setup was viewed as quick and effective by the agents and supervisors at the local and national levels. It encouraged the development of a form of creativity on the part of agents and local leaders, in terms of rules of operation, within a specific, controlled environment. Thus, the setup emerges as an opportunity for distributed intelligence, with the intelligence of the setup merging with that of the individual (Fusulier and Lannoy, 1999). In our view, this aspect is crucial to the construction and sustained implementation of organizational change (see the dialogical aspects of design described by Béguin, this volume).

Structuring the experimentation

At a methodological level, this case study allows us to identify some factors that support the structuration of the stage of transition, that is, the experimentation – which takes place notably through simulation (see Barcellini et al., this volume):

1. The *implementation of work groups* cannot be 'ordered by decree'. It is constructed with the involvement of future participants and decision-makers. As we have seen above, the configuration of group meetings, which constitute the artificial spaces used for simulations, is not totally defined in advance. In our case, the work groups were gradually built up. The construction of the experimentation relies on the social construction of the intervention. The analyses of work that had been previously carried out made it possible to share with the future participants a view of work – as a consequence, it also allowed the construction of representations about the directions that future transformations might take.

2. *Identifying and prioritizing problems.* The ergonomist provides input for the discussions that take place during meetings of the work groups, based on work situations that are validated by participants. Next, for technical reasons that are related to the intervention – i.e. all problems cannot be addressed in the meetings – it is crucial to organize these problems along hierarchical lines, in order to give the highest priority to the most sizeable malfunctions. The goal is to spark up a debate about the various logics of action involved. Criteria for quality of service have often been focused on catering for the demands of mass production (e.g. reducing production time) and taking into account the singular features of specific cases.

3. *Searching for solutions* to the problems thus mentioned does not constitute a 'technical challenge'. Indeed, problem-building is also a part of
   solution-building. The crucial constraint involves agreeing on what
   means are available for reaching these solutions. This is the object
   of a social construction between stakeholders, and questioning the
   rules of these stakeholders leads to a confrontation between logics of
   action. Therefore, searching for solutions is a dual process: one needs
   to find trade-offs to deal with the conflicts between logics of action,
   and to materialize these trade-offs in the form of technical solutions.
4. *Identifying advantages and drawbacks.* This follows on from the debates
   regarding the search for solutions. It will be necessary to identify
   criteria to assess the technical solutions that are experimented on.
5. The *experimentation stage* – a 'natural space' for simulation – must
   include a clear end date so as to integrate the time constraints of the
   project. Our experiment lasted 18 months. This may seem like a long
   time, but full-scale simulation may lead to long delays for the production of results (e.g. complaint rates of subscribers). Furthermore,
   the goal is to experiment with technical solutions and to let new
   rules of cooperation 'emerge'. This second point takes a much longer
   time to address.
6. The *evaluation of results* takes place in two stages: during the experimentation and at the end of it. The results of the experimentation
   between the departmental branches and the processing centre
   allowed us to emphasize the fact that evaluations during the experimentation aimed to adjust the solutions that had been selected. One
   final evaluation took place at the end of the experiment, in the presence
   of its participants. Its goal was an overall presentation of the results
   obtained via structural alterations. It also makes it possible to sum-
   marize the modes of cooperation involved in the experimentation.

Based on this example, we can clearly see that it is not just the objective
features of the situation that will serve as a means for action, but also the
possibility of acting upon the situation. Indeed, 'one must, here, reverse
the general opinion and agree that it is not the harshness of a situation,
nor the suffering that it may cause, that are motives to design another state
of things where things would be better for all: on the contrary, it is on the
day one can design another state of things that a new light falls upon our
pain and suffering, and we decide that they are unbearable' (Sartre, 1943,
our translation). The gradual structuring of the transition stage, as we
have described it above, allowed participants (notably, the branch supervisors) to imagine that it might be possible to improve the situation. This
constituted a crucial aspect of the problem in question. It is also the aspect
that opened up the door for solutions. The implementation of this experimental setup allowed us to obtain some time to construct solutions, time

where we were also able to implement methods and a 'conversation with the situation' (Schön, 1983). Indeed, this setup provided us with opportunities to 'make a move' (Daniellou, 2001) in order to test and assess the opportunities in the organization. It allowed us to use the results produced during the experiment to suggest a different organizational structure. In the end, based on our proposal, we attempted, on the one hand, to configure what would go on in the work situation by structuring the approach of experimentation. On the other hand, we sought to leave some room for the development of experience: new rules could be tested, but they could also be altered depending on the results they produced. The goal was to turn human experience into an object of experimentation.

Conclusion: An adaptable organization and a source of development

The shared evolutions of work organizations are very general, and they would not be enough to explain consequences to the health of workers and to the performance of production systems. This leads ergonomists to take a more precise interest in the difficulties encountered by the agents involved: on the one hand, understanding the discrepancies between real-world activity and the determinant factors of the organizational structure, and on the other hand, implementing a process aiming to manage these discrepancies. In Daniellou's (1999) view, the relationship between the activity of operators in the real world and the organizational structure can translate into an impossibility, which some workers may be confronted with when they do not have – or no longer have – the possibility to maintain a certain dynamic. In fact, this schematization of the reality of work constitutes an application of social regulation theory (Reynaud, 2003), in the sense that the individual activity is defined at the intersection of three spaces: the 'power to think', the 'power to debate' and the 'power to act' (Daniellou, 1998). The connections elaborated between these spaces mark the conditions in which this activity becomes alive. Thus, understanding the workings of the organization involves highlighting the tension that exists between these dimensions. The discrepancies between rules of control, the rules that make up the structure, and the autonomous roles, which are efficiently built up by the workers themselves (Reynaud, 2003), could be likened to the 'value' of the tension between these two poles, if one focuses on the activity of a single individual.

If the work of organizing is structured in such a way that it allows workers to reflect upon their own work (i.e. the power to think), to debate this work with their colleagues and to act upon the way this work is done, it then becomes a way for the organization to compensate for its deficiencies (i.e. to cope with variability) and for workers to build up their own health:

'work always involves an activity of constructing rules, which implies the existence of spaces to debate with one another and confront opinions. In businesses, such spaces are usually not official spaces of confrontation' (Davezies, 1993, p. 6, our translation). Hence, if this work on the organization is acknowledged and structured, it can constitute a source for the development of activity and become an organizational ability to process what has not been foreseen or prescribed (Czarniawska, 2009; Johansson Hanse and Winkel, 2008). In a sense, the experiment described above gave some structure to the work of organization by focusing on the management of major local issues, by setting up a local democracy, and by fostering a participatory approach (Kuorinka, 1997; Nagamachi, 1995; Wilson and Haines, 1997; Woods and Buckle, 2006). Furthermore, by making managers and other workers work together, the experiment led to the construction of forms of reciprocal recognition: from managers to operators, who are given a possibility to take part in the organizational change and to construct, in part, their own future activity, and from operators to managers, who are involved in the work of organization. Experiment-based change therefore allows a gradual adaptation of individuals, collectives and organizations.

Following this view, designing an organization that allows this kind of work gives this work an enabling aspect (Coutarel and Petit, 2009; Falzon, 2005; Nussbaum, 2000; Sen, 2005), in the sense that it becomes capable of coping with human and technical variability, a resilient organization (Hollnagel et al., 2006), and if need be, to question the prescriptive framework:

> The effects of ergonomic interventions can also be thought of as means to provide power to individuals and organizations, to give them additional tools that allow them to make some progress. The acquisition of skills can be viewed as a process of development of capabilities, for example, by increasing the number of options, the number of operative procedures that are available to every worker. Similarly, providing workers with spaces of liberty with respect to task goals or criteria increases their capabilities by increasing the set of available options. Finally, allowing teams to define their own collective activities increases team capabilities. (Falzon, 2005, p. 8, our translation)

This developmental view takes on its full meaning when one considers the organizational leeway that is necessary for the development of activity. Leaving some space available for the regulation of sources of variability leads us to view the organization as a means to modify the rules of work, should this be required. An adaptable organization is one that is designed like an instrument (Petit, 2005; Arnoud and Falzon, this volume) – an organization in which 'prescription is the object of genesis, and the subject therefore constructs the resources of his own action' (Béguin, 2010, p. 129, our translation). In this sense, the organization should be viewed as an instrument that is put to the service of the workers, and not as a tool that is designed, looking at work 'from the outside', to direct the activity – whose autonomy, circuits and places of decision are based upon a model of subsidiarity (Melé, 2005; Petit and Dugué, 2010; Dugué and Petit, 2013). This requires designing the organization with a human-centred approach – not just for humanistic reasons, but for the sake of effectiveness itself (Ebel, 1989), as a system that appears as a means to construct the relationship between people and objects in an independent, and no longer dual, fashion.

References

Béguin, P. (2010). De l'organisation à la prescription: plasticité, apprentissage et expérience. In Y. Clot and D. Lhuilier (Eds.), *Agir en clinique du travail* (pp. 125–139). Paris: Eres.

Caroly, S. (2010). L'activité collective et la réélaboration des règles: des enjeux pour la santé au travail. Habilitation thesis, Université Victor Segalen Bordeaux 2, France.

Clot, Y. (2008). *Travail et pouvoir d'agir.* Paris: PUF.

Coutarel, F. (2004). La prévention des troubles musculo-squelettiques en conception: quelles marges de manœuvre pour le déploiement de l'activité? Unpublished doctoral dissertation, Université Victor Segalen Bordeaux 2, France.

Coutarel, F., and Petit, J. (2009). Le réseau social dans l'intervention ergonomique: enjeux pour la conception organisationnelle. *Revue Management and Avenir,* 27, 135–151.

Czarniawska, B. (2009). *A theory of organizing.* Cheltenham, UK: Edward Elgar Publishing.

Daniellou, F. (1998). Participation, représentation, décisions dans l'intervention ergonomique. In V. Pilnière and O. Lhospital (Orgs.), *Journées de Bordeaux sur la pratique de l'ergonomie: participation, représentation, décisions dans l'intervention ergonomique* (pp. 3–16). Bordeaux: Éditions LESC.

Daniellou, F. (1999). Nouvelles formes d'organisation et santé mentale: le point de vue d'un ergonome. *Archives des Maladies Professionnelles et de Médecine du Travail,* 60(6), 529–533.

Daniellou, F. (2001). L'ergonome et les solutions. In F. Coutarel, J. Escouteloup, S. Mérin, and J. Petit (Orgs.), *L'ergonome et les solutions. Journées de Bordeaux sur la pratique de l'ergonomie* (pp. 4–16). Bordeaux: Éditions LESC.

Davezies, P. (1993). Mobilisation de la personnalité et santé au travail. Le travail d'exécution n'existe pas. *Le Mensuel de l'ANACT,* 187, 6–8.

Davezies, P. (2008). Stress, pouvoir d'agir et santé mentale. *Archives des Maladies Professionnelles et de l'Environnement*, 69(2), 195–203.

de Terssac, G. (1998). Le travail d'organisation comme facteur de performance. *Les cahiers du changement*, 3, 5–14.

de Terssac, G., and Lompré, N. (1996). Pratiques organisationnelles dans les ensembles productifs: essai d'interprétation. In J. C. Spérandio (Ed.), *L'ergonomie face aux changements technologiques et organisationnels du travail humain* (pp. 51–70). Toulouse: Octarès.

Dugué, B., and Petit, J. (2013). Ethical stakes of corporate governance, ergonomic contribution to organizational redesign aiming at subsidiarity. In *Understanding Small Enterprises (USE) Conference, Proceedings from USE to Action: Transforming Our Understanding of Small Enterprises into Practice to Create Healthy Working Lives in Healthy Businesses*, Nelson, New Zealand, pp. 80–86.

Ebel, K. H. (1989). Manning the unmanned factory. *International Labour Review*, 128(5), 535–551.

Falzon, P. (2005, September). *Ergonomie, conception et développement. Introductory conference, 40th congress of SELF*. Saint-Denis, Réunion.

Fusulier, B., and Lannoy, P. (1999). Comment "aménager par le management"? *Hermès*, 25, 181–198.

Hasle, P., and Jensen, P. L. (2006). Changing the internal health and safety organization through organizational learning and change management. *Human Factors and Ergonomics in Manufacturing*, 16(3), 269–284.

Hollnagel, E., Woods, D., and Levesson, N. (2006). *Resilience engineering: concepts and precepts*. Aldershot, UK: Ashgate Publishing Ltd.

Johansson Hanse, J., and Winkel, J. (2008). Work organisation constructs and ergonomic outcomes among European forest machine operators. *Ergonomics*, 51(7), 968–981.

Kuorinka, I. (1997). Tools and means of implementing participatory ergonomics. *International Journal of Industrial Ergonomics*, 19, 267–270.

Melé, D. (2005). Exploring the principle of subsidiarity in organizational forms. *Journal of Business Ethics*, 60(3), 293–305.

Morrison, E. W., and Milliken, F. J. (2000). Organizational silence: a barrier to change and development in a pluralistic world. *Academy of Management Review*, 25(4), 706–725.

Nagamachi, M. (1995). Requisites and practices of participatory ergonomics. *International Journal of Industrial Ergonomics*, 15(5), 371–377.

Nussbaum, M. (2000). *Women and human development: the capabilities approach*. Cambridge: Cambridge University Press.

Petit, J. (2005). Organiser la continuité du service: intervention sur l'organisation d'une Mutuelle de santé. Unpublished doctoral dissertation, Université Victor Segalen Bordeaux 2, France.

Petit, J., and Dugué, B. (2010, September). Une organisation "subsidiariste" pour prévenir des RPS. Presented at Proceedings of the 45th SELF Conference, Liège, Belgium.

Rabardel, P., and Béguin, P. (2005). Instrument mediated activity: from subject development to anthropocentric design. *Theoretical Issues in Ergonomics Science*, 6(5), 429–461.

Rabardel, P., and Pastré, P. (2005). *Modèles du sujet pour la conception*. Toulouse: Octarès.

Reynaud, J. D. (2003). Régulation de contrôle, régulation autonome, régulation conjointe. In G. de Terssac (Ed.), *La théorie de la régulation sociale de Jean-Daniel Reynaud* (pp. 103–113). Paris: La Découverte.

Sartre, J. P. (1943). *L'être et le néant*. Paris: Gallimard.

Schön, D. A. (1983). *The reflexive practitioner: how professionals think in action*. New York: Basic Books.

Sen, A. (2005). Human rights and capabilities. *Journal of Human Development*, 6(2), 151–165.

Vézina, N. (2010). TMS ailleurs. Prévention des TMS au Québec. *Archives des Maladies Professionnelles et de l'Environnement*, 71(3), 426–430.

Wilson, J. R., and Haines, H. M. (1997). Participatory ergonomics. In G. Salvendy (Ed.), *Handbook of human factors and ergonomics* (pp. 490–513). Chichester, UK: Wiley & Sons.

Wisner, A. (1972). *Le diagnostic en ergonomie ou le choix des modèles opérants en situation réelle de travail*. Report 28. Paris: Laboratoire de Physiologie du Travail et D'ergonomie, CNAM.

Woods, V., and Buckle, P. (2006). Musculoskeletal ill health amongst cleaners and recommendations for work organisational change. *Journal of Industrial Ergonomics*, 36(1), 61–72.

chapter ten

The design of instruments viewed as a dialogical process of mutual learning

Pascal Béguin

Contents

The International Ergonomics Association defines ergonomics as 'the profession that applies theory, principles, data and methods to design in order to optimize human well-being and overall system performance' (http://www.iea.cc/ergonomics). This orientation of its program has led to diverse work within activity ergonomics. We will describe a brief panorama of this work, and arrive at the idea that one cannot limit oneself to designing artifacts. Design is a process of joint development of artifacts and of the activity of those who use them. The goal of this text is to propose a dialogical approach of mutual learning between workers and designers, aiming to address the challenge of a developmental approach to design.

Crystallization, plasticity and development

The concept of activity, as well as the understanding that we have of the processes of design, is not set in stone in the discipline. Over time, this knowledge has evolved, leading to the birth of various perspectives. Three perspectives can be distinguished, which we will term crystallization, plasticity and development.

Crystallization

The central idea behind this first view (which is the oldest) is that every technical device, every artifact, 'crystallizes' knowledge, a representation, and in the broadest sense a *model* of the user and his or her activity. However, once they are crystallized in the artifact, these models may become a source of difficulty (or even exclusion) for people if they are false or insufficient. For example, planning a staircase to access premises relies on a representation of valid subjects that, once it is crystallized in the artifact, imposes itself upon everyone. With the risk of excluding people in a wheelchair: these people will be unable to access the upper floors. This is a general feature of design: the software 'sets' in the artifact a psychological model of the user (Carroll, 1989; Bannon, 1991).

This can be generalized. Any technical system integrates, materializes and conveys numerous choices made by its designers regarding the nature of the work to be done, but also choices that are of a social, economic and political nature (Freyssenet, 1990). Yet, ergonomists will note that these choices are often made based upon an insufficient knowledge of the workings of man, and of the conditions in which the work is to be done.

During design, ergonomists must then offer and diffuse better models of the workings of humans and of their activity when faced with technical objects, and turn these models into resources for design. Indeed, a large part of earlier work in ergonomics aimed to provide better data about humans (e.g. anthropometric data). Methods of ergonomic work analysis (EWA) were fundamentally developed toward this goal: to construct 'operant models', that is, according to Wisner (1972), 'models of the situation that are representative of some essential aspects of the real world, and are likely to lead to effective solutions' (our translation). The issue is, as Maline (1994) might say, to 'transform the representations' of designers (our translation). Yet, EWA makes it possible, first and foremost, to objectify a situation that already exists (Theureau and Pinsky, 1984). Yet, in design, one must project oneself in the future – leading to the idea that one should anticipate 'future activity' in order to model it.

Plasticity

However, this wish to anticipate human activity leads to debates that are epistemological (how is it possible to know this future activity?) and, above all, ontological (regarding the nature of activity). Indeed, a wide range of empirical and theoretical work has shown that there will always be a difference between activity as it can be apprehended and modelled for design and activity as it will be, in fact, deployed in a given situation.

Activity is guided by concrete situations. But these situations are constantly evolving because of industrial diversity and industrial variability (Daniellou et al., 1983). Indeed, in professional situations, workers encounter unexpected events, resistances that are related to industrial variability – tool malfunctions, instability of the matter that must be transformed, absence of a colleague, etc. Hence, whatever efforts are made to anticipate activity, the performance of action will never completely match what had been planned out. Workers, then, are expected to exhibit an 'intelligence of the task' (de Montmollin, 1986), to adjust to these events, to take into account the contingencies of the situation, for example, by acting at the right time and by making use of favourable circumstances. In short, human activity is 'situated' (Wisner, 1995).

This situated view of activity (in the sense of the theoretical framework of situated action; see Suchman, 1987) leads to redefining the goals of ergonomists in design. The goal is not (just) to construct a more grounded representation of humans and the way they function. One must (also) design systems that are flexible enough to leave some degrees of freedom to the activity in real life, in terms of both production effectiveness and the health of workers (to achieve production goals without endangering one's health).

Let us note that this orientation leads to a research program whose goals are twofold: both technological (what are the properties that systems should have in order to be plastic and afford leeway?) and methodological (how can one influence design to take into account industrial variability and diversity?). The work carried out in activity ergonomics has led to an original position: the goal is to contribute to the specification of 'spaces of future activities', or of possible activities (Daniellou, 2004). This author provides an example: providing workers with a printer allows them to rely on a printout when required. But in the absence of a printer, the only possibility is to rely on the screen. In methodological terms, the goal is then to make an inventory of 'typical situations of action' (i.e. the diversity of probable contexts of action) in order to examine whether the worker will have access to the required leeway – in terms of resources and constraints – allowing him to cope with the diversity of situations in the real world – in terms of consequences for oneself and for production goals.

Development

The approach described above leads to a crucial idea: the anticipations and internal references constructed in the act of designing are not enough. One must take into account activity in real life.

However, in the situated approach of plasticity, the operator acts in order to cope with events. This interpretation, which focuses on extrinsic aspects (tasks, tool malfunctions, the absence of a colleague, etc.), is obviously not wrong. However, it is insufficient. The work carried out by operators also has origins in the intrinsic aspects that are related to the worker himself and to his activity. Let us emphasize three ideas.

The first idea is that every artifact (machines, tools or production processes) is ultimately implemented by workers, who will mobilize their ways of doing and thinking (skills, operative concepts, etc.), to ensure that it works well. Yet, anthropology of technology (see e.g. Geslin, 2001) has abundantly demonstrated that there is no such thing as a void of technique (in the etymological sense of 'effective know-how') but, on the contrary, that there exist some antecedent cognitive and cultural constructs. This is true even when the artifact is radically new to the social and cultural environment in which the technological innovation is introduced.

The second idea is that these preexisting ways of doing and thinking things will often be put into motion by the technical novelty itself. The introduction of a technical novelty in a given situation often makes it possible to solve old problems. However, it also changes the nature of the task and creates new problems for which new forms of action will be required. Hence, the question emerges of the genesis of a new activity in the face of novelty and of the designed object.

The third idea is that if one tries to analyze these geneses – the processes whereby operators appropriate themselves a technical novelty and turn it into a resource for their actions – one can see that they relate to two different, distinct forms: either the worker develops new techniques based on those that are already available to him, or he adapts, alters or transforms devices to make them compatible with his own constructions. Indeed, this is the main result of the work carried out regarding 'instrumental geneses' (Béguin and Rabardel, 2000; Bourmaud, this volume). During the appropriation of a technical device, one can observe either an instrumentation – workers modify their activity to match it to the device – or an instrumentalization – that is, matching the novelty to the activity: workers adapt, alter, reinterpret or even transform devices (either temporarily or durably) to match them to their own ways of thinking and doing things. In work situations, and when faced with radically new situations, one can obviously observe large-scale processes of appropriation, often connecting together instrumentation and instrumentalization. These processes sometimes take place over many years.

Diversity and unity of the connections between design and work

The three perspectives we have presented above pose, in fact, two questions. The first relates to the object that is to be designed: is it the technical design, the leeway or the activity? The second relates to the unit of action on which the intervention of ergonomists in design can be based.

Designing a coupling

The concepts of crystallization, plasticity and development exhibit many differences. But beyond this diversity, all three argue the same fact: one should take into account at the same time the features of technical systems, on the one hand, and the activity of workers, on the other hand.

Indeed, crystallization emphasizes the fact that the work activity must be modelled at the same time as one specifies technical tools and devices. Plasticity argues that the effectiveness of devices does not rely exclusively on decisions made in design departments, but also on the situated activity. And development indicates that activity develops over the course of the development of a tool. Therefore, it is a coupling – the systematic organization of two entities – that constitutes the object that is being designed to the ergonomist in the midst of an intervention.

The instrumental approach – mentioned above – has particularly underlined this point, by asserting that a distinction should be made between the artifact and the instrument (Béguin, 2006; Bourmaud, this volume). The artifact is a manufactured object in its material and symbolic aspects. The instrument, on the other hand, is a composite entity that comprises both an artifact and a component related to action. It is the association of these two, organized in a system, that forms the instrument. For example, a pen is an artifact that does not in itself form an instrument. In order for it to be an instrument, it must be associated with organized forms of action (that are, indeed, the object of a long learning process in children). It is the association of the two that allows writing.

Therefore, the instrument can be defined as a bipolar entity, connecting two components: a 'human face', which comes from the subject (the worker, the user), and an 'artifactual face' (an artifact, part of an artifact or a system of artifacts), which is of a material or symbolic nature.

One of the interesting aspects of this conceptualization is that it argues that even if an artifact is very well designed, the instrument is by no means finalized when it leaves the design departments. In any case, the 'living' instrument, the one that is in fact implemented, implies that a human being, either a worker or a user, should associate it with part of oneself. However, no one can replace this 'part of oneself'. This definition of design thus suggests that designers and users contribute to design based on their own competence and diversity.

Mutual learning and dialogues

If, from the point of view of methods and approaches for design, one wishes to follow the preceding stance (that is, workers and designers contribute to the design based on their diversity), it is not enough to focus on the coupling between the subject and the design object (as in all the works we have mentioned thus far). One must take into account not just the activity of workers, but also that of designers. Indeed, behind the artifact is the activity of designers, and these must have a status in the process.

In the sense that designers and workers contribute to design based on their diversity and own competence, one must, from the point of view of methods, focus on the dynamics of the exchanges between workers and designers in order to facilitate and provide tools supporting them. It is in this view that we will speak of a *dialogical approach of mutual learning* (Béguin, 2003; Barcellini et al., this volume). Two elements characterize this approach:

- As is argued by the famous metaphor of a 'conversation with the situation' (Schön, 1983), each designer carries out, over the course of one's activity, some learning. The designer, aiming for a goal, projects ideas and knowledge. But the situation 'responds' and is a source of 'surprise', because it presents some unexpected forms of resistance that are a source of novelty, leading to learning and reorganizations. However, since design is a collective process, the other stakeholders of the process also respond to and surprise the designer. In this context, the result of a designer's work is at best a hypothesis, steering the learning of other people. But this learning may be possible or impossible, leading – depending on the case – to validate, but also sometimes to refute, or more simply to put in motion, the initial hypothesis. This response may then lead to starting up a new cycle of learning, this time on the part of the person having formulated the initial hypothesis. Thus, this approach leads to an interesting model, if one extends it to interactions between designers and workers. In this model, one switches from the idea of the appropriation of a novelty to that of learning and of a confrontation between different forms of knowledge. In this model, a novelty designed by the designers may lead to learning on the part of workers. But this model also posits that designers can also learn from workers. There are therefore mutual processes of learning that should be supported and supervised. In this context, the productions of designers should be viewed as 'instrumental hypotheses'.

- Following such an approach, the instrument develops over the course of dialogues between workers and designers. However, this concept of dialogue should not be understood only in the narrow sense of verbal communication. It is a process with a dialogical structure: the result of the work of one person is put back to work within the activity of another, leading to a response. However, the vectors of this dialogue can be, for example, a blueprint, a mock-up or a prototype. On this particular point, we join here the issue of 'intermediary objects' of design. This concept of intermediary object posits that the objects of design are media, used for representation and communication between the stakeholders of design. In our own words, the intermediary objects constitute the vectors of the instrumental hypotheses. Hence, they make it possible to focus the exchanges between stakeholders on the function of the artifact and on its expected use by workers. These hypotheses may then be validated or questioned, when they are subsequently confronted with the activity of the workers. Dialogical processes may, then, include discourse, relying on language. But they fundamentally involve action, specifically in a confrontation of what is designed with the resistances of the real world.

Organizing dialogical design

This dialogical model provides a very different vision from the classical approach of engineering, where design is viewed as a change of state during which problems must find a solution. In a dialogical view, design appears not to have a true starting or finishing point. It is, instead, a cyclical process where the results of the work of one person, designer or worker fertilize the work of another, and where there is probably no such thing as a last word.

This view opens up several possibilities for organizing design. By drawing on the concept of instrumental hypothesis that we have just introduced, I propose three directions that concern three different moments in time.

Objectifying the instrumental hypotheses of workers

This first path consists in instituting the results of the work of operators as an instrumental hypothesis, a source for the activity of the designer. The activity of the user, in this view, therefore temporarily gains 'first ground'. Let us take the example of the design of a machine to fold and insert mail into envelopes.

This study was carried out within a national centre in charge of sending the administrative correspondence of a ministry (Béguin et al., 1998).

It was carried out following a request from management, and after the arrival of a new machine. This machine was entirely automated, was much faster than earlier machines, and allowed inserting more than five documents in a single envelope. Unfortunately, the workers proved unable to bring out the produced mail for several months. Work analysis made it possible to identify the causes of these difficulties. To set a machine for folding and inserting the documents in the envelopes, it is necessary to construct a representation of the kinetics of the paper while the machine is in operation, in order to see how the paper will be warped. Indeed, it is impossible to define proper settings for the machine and to avoid paper jams without being able to do this. However, because the new machine included a hood, it was impossible to collect this information regarding paper warps (in spite of the large amount of information that was available on computer screens). However, work analysis also showed that the older machines, which had been used before the introduction of the new automated machine, had been greatly altered by the workers. They had altered the hoods to add some Plexiglas windows, thus facilitating the collection of significant information. The alterations that had been made were therefore analyzed from the point of view of their functions. These data served as a basis for the construction of functional specifications for the design of a new automated machine.

In this example, the alterations that workers had made on the machines were viewed as instrumental hypotheses. They were not implemented 'as is' by the designers. It is the identification of functions and user needs that is at the source of the transformations carried out by the workers that is of interest to us. Technical possibilities may allow us to provide novel solutions, which the workers may not have identified yet. Objectifying the workers' instrumental hypotheses therefore requires specific work on the part of the ergonomist, relying on his or her ability to take into account the level of uncertainty of the design process, and the corresponding work of designers.

Designing instrumental hypotheses

In this second view, the activity of the designer is (temporarily) in the forefront. This activity consists in specifying the artifactual side of the instrumental hypotheses, by supposing, during the design of the artifact, that its properties may be altered by the workers based on the development of their own activity.

Our example will be the LENS software program, used to manage electronic mail. This piece of software has led to an entire series of studies. It is a software program that was originally designed as an 'intelligent' agent. Mackay (1988) has shown that the workers alter the functions afforded by the software: they wish to be notified of the arrival of new messages, and to be able to consult these messages. Therefore, the users develop new functions. For example, they use the system as an assistant that will archive the messages in an adequate area, and not as a filter. Above all, one of the advantages of this software is that it allows each user to construct his or her own filter, taking into account personal needs (de Keyser, 1988).

Here, we return to a technological orientation – that of transformable systems. One of the important roles played by ergonomists in designing these artifacts is grounded in the fact that they must define the properties of the artifact and the activity corresponding to those properties. Indeed, focusing on the artifact itself, it will be necessary to specify the levels in which the system can be modified (e.g. systems that cannot be modified, systems that can be modified and adapted within the limits and views expressed by the designer, and systems that can be modified in new ways, from a functional point of view) (Randell, 2003). However, this implies identifying the various kinds of practices that correspond to these levels (for example, choosing between options that have been determined in advance, or designing new behaviours for the artifact based on existing elements).

Furthermore, the issue of transformable systems must not be borne by the ergonomist only at the level of the instrument. Modifying technical systems requires resources. Those may be, obviously, cognitive resources (e.g. the existence of an instruction booklet, the possibility of exchanging information with the designers when required, etc.), but also time resources. Such an activity of 'sustaining design' in a work situation requires time. It must therefore be allowed, not just by the artifact, but also by the simulation.

Engineering resonance between the instrumental hypotheses of workers and designers during design

In the two directions mentioned above, the activity of the designers and that of the workers are asynchronous. The third direction is to organize synchronous work. The specific feature of this third way is that the dialogical exchanges between designers and users serve as a driving force for design, as we will see with an example regarding the design of an alarm aiming to prevent runaway reactions in SEVESO class chemical plants (Béguin, 2003).

We are in a production unit in the field of high-precision chemistry. The product being manufactured is highly explosive, and can exist in either one of three states. In the cold state, it thickens and hardens, leading to 'supercooling'. When it heats up, it becomes a liquid, which is its ideal state. But when it is too hot, it produces a highly explosive gas, leading to a runaway reaction. Runaway reactions are the primary cause of human casualties in chemical production. Workers seem to remain on site until the explosion occurs. This is why the engineers have designed a safety system. This system is an alarm whose goal is to predict a mean remaining time (MRT) before the explosion occurs. First, a detection algorithm was developed and tested in an experimental situation. It was then suggested that a prototype of the artifact be introduced in a tryout situation. This prototype displayed (1) the mean remaining time before an explosion and (2) very precise information regarding the temperature of the product, to within a hundredth of a degree Celsius.

The results of the introduction of the prototype on the experimental site speak for themselves. They show that the workers gradually consulted the alarm more and more. Thus, in the first trials, they consulted the interface for only 1.7 per cent of the time devoted to the supervision of the production process. However, 3 months later, they consulted it for 31.5 per cent of the time (see Table 10.1). Therefore, the workers took ownership of the system, which is a priori a good thing.

However, a more detailed analysis of information collection (based on gaze direction) showed that the workers use the alarm instead of the thermometers that were previously available (see Table 10.1). Thus, the workers collect mainly the information that the device gives about temperature, and not the time remaining before the explosion (MRT), although this is the whole point of the system. In short, in the activity of the workers, the alarm takes on the status of a thermometer.

However, this status is unacceptable from the point of view of the designers. First, European regulations (NE 31) require that 'safety instrumented systems' should be separate from the other available tools. Whereas the alarm was initially designed as a safety instrumented system, it takes on the status, in the activity of the workers, of an 'aid to

Table 10.1 Comparison between the Evolution of the Time Devoted to Information Collection Using the Prototype and Using the Thermometers That Were Previously Available on Site[a] between the First, Second and Third Trial Sessions for the Prototype

	First trial session	Second trial session	Third trial session
Duration of the collection of information on the prototype	1.7%	8.1%	31.5%
Duration of the collection of information on the temperature indicators that were previously available	27.3%	24.5%	4.2%

[a] In per cent of the total duration of work.

process supervision'. Furthermore, what role does this instrument take on with respect to the major hazard of runaway reactions, whose prevention is the main goal of the project? The fact that workers took such ownership of the system placed the designers in a very uncomfortable situation. From their point of view, it even meant that the project had failed.

Starting with the idea that unexpectedly taking ownership of the system constituted a response on the part of the users, we carried out an analysis of work to understand this response. This analysis showed that the evolution of the function of the artifact that took place within the activity of the workers found its source in the strategies these workers used to control the production process. Indeed, the workers controlled the process by maintaining it at the lowest possible temperature threshold. This strategy 'drives them away' from the major hazard of a runaway reaction (which occurs when the temperature threshold is set high). However, running the process at a low temperature is also hazardous: if the product becomes too cold, it may crystallize and harden. To use the expression of the workers themselves, this is an 'everyday hazard'*. The designers took into account this need of the workers, and altered the artifact accordingly: in addition to the initial analogical display of temperature, they included the display of a curve, making it possible to track the history of the evolutions of temperature. Indeed, the curve makes it possible to interpret trends in the product's thermal kinetics. This emerged, from the analysis of work activity, as a variable used by the workers to prevent these everyday hazards.

In addition, the work analysis showed that the workers spent a lot of their time trying to steer clear from the major hazard of runaway reactions, that is, to produce while staying as far away as possible from high temperatures. However, runaway reactions might still appear following, for example, a breakdown or damages to the equipment. Once again, to use the workers' own words, they would then have to 'cope with the unknown' – hence the idea of designing an instrument that would allow workers to apprehend the concrete conditions of a runaway reaction. The alarm was able to achieve this, since its design was based on a model of runaway reactions. Therefore, the prototype was modified so as to be able to simulate the temporal dynamics of such major events (to do this, the product being manufactured was replaced with an inert liquid). Three simulations were conducted that differed from one another in terms of safety procedures that had been planned for that particular site. They showed that, in two of the cases, the workers would have failed

---

* As it expands, the product may cause some of the equipment, which is made of glass, to break. Furthermore, it will be necessary to heat the product up later on, which may be hazardous.

to cope with the hazard – in one case, because of the organizational conditions of work, and in the other, because of the architecture of the site. Following this, an additional worker was hired and the architecture of the site was modified.

In this example, the characteristics of the first version of the artifact corresponded to the instrumental hypothesis of the designers. However, the workers produced a (creative) response that modified its function and its meaning. The designers took this into account. However, the response of the workers did not draw the dialogue to a close. Instead, it generated another response on the part of the designers, who developed an artifact whose aim was, in a sense, to bring the workers on their own 'home field' – by allowing them to experience the concrete conditions of a runaway reaction. This led to broader alterations, regarding the site's organizational and architectural conditions.

Constructing the dialogical frames of design

In this chapter, we have proposed a dialogical model of the mutual learning that occurs in design. One of the stakes of this approach is to locate, on a single scene, within a single frame of action, the heterogeneous logics and positions of the designers and workers, for them to work together.

However, the frame of a dialogical design is not limited to its temporal aspects. It also lies in the power games between the stakeholders of the design project. A dialogical model of design tends to make these stakeholders all symmetrical, and to focus on their knowledge, but also tends to abandon the issues related to power plays. In the sense that design is characterized by heterogeneous points of view, workers and designers can legitimately disagree with each other. However, these disagreements between stakeholders may be dealt with following two contrary approaches:

- One pathway is conflict, e.g. through the exercise of authority or the exclusion of some of the stakeholders, whose goals, motivations and decision criteria appear to be overly divergent or are viewed as insignificant.
- The second pathway is, precisely, design. Disagreements serve as a driver to modify the characteristics of the object being designed. Criteria are changed, specifications are adjusted, and goals are redefined so the solution can be viewed as acceptable within the group.

The difference between these two opposing pathways is that, in conflict, disagreements are resolved through a confrontation between stakeholders. The strongest wins. In design, however, the complexity of the exchanges can be attributed to the complexity of the real world. Between these two pathways, there is a shift in balance. In the first case, difficulties are dealt with by suppressing the diversity of points of view within the group, while the complexity of the real world fades into the background. In the second case, the very purpose of design is to solve difficulties by managing the complexity of the real world, while being mindful of the diversity of points of view within the group.

The existence of these two paths raises the issue of the relationship between knowledge and power. Foucault (2004), who strongly emphasized the importance of this relationship, set a distinction between two kinds of devices: 'normation' devices and 'normalization' devices. Normation is characterized by the fact that knowledge turns into power. It then becomes the norm, and those who do not conform to it turn to abnormality. The second device, normalization, consists in the gradual construction of curves for the development of knowledge, in order to achieve a local institution of normality.

In many ways, the proposals made in this chapter aim to guide design toward a less normative form in order to achieve the local institution of normality through dialogical forms of design. One could also argue that such an approach is favourable to the health of workers. Indeed, Canguilhem (1966) has argued that the 'healthy man' is a man who is not subjected to the constraints of one's environment, but who is capable of changing them to assert one's own norms (for example, professional norms) and life project.

The fact remains, however, that normation and normalization are two sociocognitive frames that are built before the intervention of ergonomists. The ergonomist's intervention takes place in a social context that precedes it*. But in any case, the role played by an ergonomist is never neutral. This joins the position of Daniellou and Garrigou (1993): the ergonomist plays a very active part in 'reframing' the exchanges between stakeholders. The ergonomist may even be, I believe, a 'guardian' of this frame. But such a formulation implies that we must clearly see the social and axiological dimensions of the frame. Both dimensions constitute an obligatory background to dialogism in design – and as a consequence, to developmental approaches in ergonomics.

---

* Indeed, this is a dimension that imposes itself on the ergonomist, and that is probably all the more salient when one considers sectors with low added value, involving workers with low qualifications.

References

Bannon, L. (1991). From human factors to human actors. In J. Greenbaum and M. Kyng (Eds.), *Design at work, cooperative design of computer systems* (pp. 25–45). Hillsdale, NJ: Lawrence Erlbaum Associates.

Béguin, P. (2003). Design as a mutual learning process between users and designers. *Interacting with Computers*, 15(5), 709–730.

Béguin, P. (2006). In search of a unit of analysis for designing instruments. *Artefact*, 1(1), 32–38.

Béguin, P., Millanvoye, M., and Cottura, R. (1998). *Analyse ergonomique dans un atelier de mise sous pli*. Research report. Paris: CNAM, Laboratoire d'Ergonomie et Neurosciences du Travail.

Béguin, P., and Rabardel, P. (2000). Designing for instrument mediated activity. *Scandinavian Journal of information Systems*, 12, 173–190.

Canguilhem, G. (1966). *Le normal et le pathologique*. Paris: PUF.

Carroll, J. M. (1989). Taking artifact seriously. In S. Maass and H. Oberquelle (Eds.), *Software-Ergonomie '89* (pp. 36–50). Stuttgart: Tentner.

Daniellou, F. (2004). L'ergonomie dans la conduite de projets de conception de système de travail. In P. Falzon (Ed.), *Ergonomie* (pp. 359–374), Paris: PUF.

Daniellou, F., and Garrigou, A. (1993). La mise en œuvre des représentations des situations passées et des situations futures dans la participation des opérateurs à la conception. In D. Dubois, P. Rabardel, and A. Weil-Fassina (Eds.), *Représentations pour l'action* (pp. 295–309). Toulouse: Octarès.

Daniellou, F., Laville, A., and Teiger, C. (1983). Fiction et réalité du travail ouvrier. *Cahiers Français de la Documentation Pédagogique*, 209, 39–45.

de Keyser, V. (1988). De la contingence à la complexité. L'évolution des idées dans l'étude des processus continus. *Le Travail Humain*, 51, 1–18.

de Montmollin, M. (1986). *L'intelligence de la tâche*. Bern: Peter Lang.

Foucault, M. (2004). *Sécurité, Territoire, Population, Cours au Collège de France, 1977–1978*. Paris: Hautes Etudes/Gallimard/Seuil.

Freyssenet, M. (1990). *Les techniques productives sont-elles prescriptives? L'exemple des systèmes experts en entreprise*. Cahiers du GIP Mutations Industrielles. Paris.

Geslin, P. (2001). *L'apprentissage des mondes. Une anthropologie appliquée aux transferts de technologies*. Paris: Maison des Sciences de l'Homme.

Mackay, W. (1988). *More than just a communication system: diversity in the use of electronic mail*. Working paper. Sloan School of Management, MIT. http://dl.acm.org/citation.cfm?id=62293.

Maline, J. (1994). *Simuler le travail, une aide à la conduite de projet*. Paris: ANACT.

Randell, R. (2003). User customisation of medical devices: the reality and the possibilities. *Cognition Technology and Work*, (5), 163–170.

Schön, D. (1983). *The reflective practitioner. How professionals think in action*. New York: Basic Books.

Suchman, L. (1987). *Plans and situated actions*. Cambridge: Cambridge University Press.

Theureau, J., and Pinsky, L. (1984). Paradoxe de l'ergonomie de conception et logiciel informatique. *Revue des Conditions de Travail*, 9.

Wisner, A. (1972). Diagnosis in ergonomics or the choice of operating models in field research. *Ergonomics*, 15, 601–620.

Wisner, A. (1995). Understanding problem building: ergonomic work analysis. *Ergonomics*, 38(8), 1542–1583.

chapter eleven

From use analysis to the design of artifacts

The development of instruments

Gaëtan Bourmaud

Contents

Various works in ergonomics have attempted to develop and implement conceptual and practical frameworks for designing artifacts. The relations that are developed and maintained between people and technology can, then, be viewed in different ways. The goal of this chapter, which focuses specifically on the use of artifacts and on their integration in activity, is to propose a framework for constructive ergonomics to think and act in design.

One of the main frameworks for design, traditionally called human–computer interaction (HCI), structures itself around the following question:

how can we design artifacts that can be used easily, efficiently, comfortably, etc. by their users? The stakes of design then aim toward artifacts whose properties are well suited to the characteristics of (future) users. In such a framework, design appears to be guided, first, by the presumed and antici-pated use that users should make of it. What is being sought after is a bet-ter fit between user needs, which are themselves an object for analysis, and the functions of these artifacts, in order to perform tasks that have been determined beforehand, and which are viewed as well known and stable. Furthermore, the artifacts are understood through their *interface* aspect, which designates the part that is *at play* in the relationship with the users; this is almost a bijection. Designers speak of HCI, and the goal of design is to ensure, as much as possible, quality in this interaction. Usefulness and usability are emphasized as criteria that can guide design choices – along with the other criteria cited above, such as efficiency, comfort, etc. – so that users can genuinely and easily take ownership of the artifacts.

Some other frameworks exist, and one of those – the instrumental approach – considers that the approach of designing artifacts by focusing on the dialogue or exchanges with the interface is overly restrictive. In this approach, artifacts are regarded as technical proposals, which may – or may not – become means for action in the activity of workers. In this chapter, we will use this term rather than that of users, because it relates to the idea that the use of artifacts is guided by a goal that is both distinct from and of a higher order than the mere use of these artifacts (Daniellou and Rabardel, 2005). It is the specific, situated uses that the workers will make of these artifacts that will allow this potential to materialize. At this point, these artifacts become instruments.

However, it would be a pity to view the contribution of this approach only from the point of view of understanding or analyzing the relation-ship between workers and artifacts. Indeed, rather than just viewing this relationship as a fact, the goal is to consider the instrumental approach as a means for the ergonomic intervention itself. From this point of view, it seems relevant to attempt, as a goal for ergonomic interventions, to design artifacts that might become instruments.

Thus, analyzing the emergence and development of instruments, on the one hand, and integrating the product of this analysis in design pro-cesses, on the other hand, can be viewed as two relevant directions to propose a constructive view focusing on artifacts and their design.

From artifacts to instruments

The concept of instrument was originally proposed by Rabardel (Rabardel, 1995, 2001; Rabardel and Bourmaud, 2003) and further used by other authors (Béguin, 2003; Béguin, this volume; Bourmaud, 2006; Folcher,

2003). In order to describe this approach, it is useful to describe its basic principles (Folcher and Rabardel, 2004; Rabardel and Waern, 2003).

The first principle of the instrumental approach relates to a proposal made by Norman (1991) to look at instruments from the 'personal view' rather than the 'system view'. The system view suggests examining the worker, the task and the artifact as constituting a system whose performance is improved and amplified in comparison to any of these three components considered on its own. In this view, the worker alone is considered a component whose ability is limited – perhaps too limited. Conversely, the personal view argues that the artifact changes the nature of the task and of the worker. The task is then modified and restructured, leading to an impact on the worker himself. The second principle of the instrumental approach asserts that the worker and the artifact should not be seen as symmetrical entities of an interaction within a given system. Instead, they are in an asymmetrical relationship where the interaction is the work of the worker himself and is, consequently, intentional. This principle relates to activity theories, since it considers that artifacts take on a specific place as mediators of the activity of the workers. This assertion constitutes the third principle and rests on the work of Vygotsky (1931–1978) and on what was subsequently known as *Activity Theory*. Another principle is that artifacts are not just objects with a specific form and specific physical properties. Indeed, artifacts carry within themselves some social and cultural features (Cole, 1996; Leontiev, 1978; Wertsch, 1998). Artifacts are part of a history that goes beyond that of a single worker, and includes shared contributions. Subsequently, and as a fifth principle, these same authors have demonstrated that artifacts are objects that undergo development. This developmental view makes it possible to apprehend how the worker takes ownership of the artifact, viewing this as a process that is obviously a gradual construction, both in each situation and in the worker's own personal history. The sixth and final principle is grounded in the theoretical approach of situated action (Suchman, 1987): action is driven toward a goal, and depends on the social circumstances and material resources that are used. Regarding the specific focus of this chapter, the situations encountered by the worker influence the activity in a crucial manner. An activity that is mediated by artifacts is therefore always a situated activity.

Hence, the instrumental approach proves to be particularly useful in order to account for the relationship between workers and artifacts (Kaptelinin and Nardi, 2006). It is a particularly elaborate conceptualization, which we will describe in greater detail below.

Instrument as a coupling between an artifact and a schema

The concept of instrument, as a proposition, integrates a fully hybrid character: it is both in part artifactual and subjective (in the sense of what can be done by the subject as part of artifact use). It is this coupling between the artifact and the schema as defined below, which is carried out by the worker himself, in a specific situation and in pursuit of a specific goal, that makes it possible to determine the instrument.

The artifact may take on various forms and may be produced by the worker himself, or by other people for the worker's benefit. In some cases, only a part of the artifact will turn into an instrument – such as in the case of using only some of its functions or some of its display screens.

The concept of schema is derived from the works of Piaget (1952). In the end, it refers to the concept of 'use' as we have used it thus far. Schemas constitute the means whereby workers are able to apprehend the situations and objects to which they are confronted as part of their interactions with their environment. Schemas respond to two different processes:

- Accommodation: Workers may draw from the schemas that they have constructed throughout the course of their own personal histories, and transform or reorganize these schemas in order to respond to the new situations they encounter. Hence, for example, to use new artifacts, workers will transform their own schemas by operating a process of differentiation.
- Assimilation: The worker may thus apply similar schemas to artifacts despite their being very different from one another, and implement them in a suitable manner as part of a process of generalization.

Schemas present another dual feature. They are also of a private and a social nature. Throughout their own personal histories, workers will construct their own schemas (through assimilation and accommodation). Yet, each of them is not in complete isolation, and other workers may take part through practices aiming to share schemas and to hand them down, for example, between peers belonging to the same community (Bourmaud, 2006) or to a work collective.

The instrument as a construct

The example below may help us introduce the proposal made here. A walking stick purchased in a store presents some intrinsic features that guide the use that can be made of it in a ramble: its adjustable height, its moulded handle, its strap, the pick that is fixed at the end, etc. all strongly guide how it can be manipulated and used for support in walking paths. However, all of us have experienced forms of use that are less expected for this stick, and yet very practical: using it to pick a fruit by reaching out, extending the arm and swiping at it can prove quite an enjoyable experience. The concept of affordance (Gibson, 1979; Norman, 1988) is useful here: it refers to the perceptible attributes and properties of artifacts that make a specific type of action possible.

However, this latter use of the walking stick can be seen in two different ways, depending on what value one ascribes to it. One can believe that it 'wasn't designed to be used like that'; it is a form of use that is of lesser value, since it was not planned or anticipated by the designers, leading to what has been called a catachresis (Faverge, 1977). Alternately, from a constructive perspective, one can see that such uses are an obvious marker of creativity on the part of the user.

This idea is not a new one, and there are many works that aim, following the approach of Activity Theory, to show that the means to mediate activity are not provided to workers at once (Bannon and Bodker, 1991; Kaptelinin and Kuutti, 1999; Kaptelinin and Nardi, 2006; Wertsch, 1998). This reconceptualization of unexpected uses of artifacts thus provides a fundamentally different approach: the genesis of an instrument takes place through the worker himself. This view is particularly meaningful for the constructive point of view of the instrumental approach, and applies just as well to artifacts that rely on sophisticated technology, as the example at the end of this chapter will show.

Two processes seem to be involved in this *instrumental genesis*, which are characterized by their orientation. These are the processes of *instrumentalization* and *instrumentation*, which both contribute to the development of the artifact as well as that of the worker.

The process of instrumentalization

Instrumentalization affects the artifact. It is a process that can be considered an enrichment of the properties of the artifact by the worker. This process includes all that bears upon the selection, the grouping, the attribution of properties and functions, or even the transformation of the artifact. Here, we can find, for example, the possibility afforded by some software programs to personalize the interface. But this process goes far beyond that, since it includes and goes beyond the freedom of use that had been left by the designers for the workers.

In some cases, instrumentalization does not involve any physical transformation of the artifact. The example given by Faverge (1977) is quite well known: the wrench from a toolbox may be used without any alteration whatsoever to strike objects as with a hammer. This attribution of functions may be either temporary and related to a specific action or longer lasting. However, the artifact is very often altered. The use that is made of it leads to adapting the properties of the artifact to the situation encountered. Alternately, the artifact might have been designed by workers themselves and undergone, either initially or subsequently, this process of instrumentalization – as in the case of making a walking stick out of a broken branch that one has found lying on the ground.

Instrumentalization therefore is characterized by the emergence and evolution of the functions of the artifact and by the evolutions of the functions of the artifact that are carried out by the worker himself.

The process of instrumentation

The process of instrumentation affects the schema or schemas of use, and deals with their emergence or evolution. Instrumentation is therefore oriented toward the worker. It is a process of development of instruments that plays out at a level that is internal to the worker; it is the result of a construction that is the worker's own doing.

Rabardel (1995, p. 143) notes 'the gradual discovery of the (intrinsic) properties of artifacts by the subjects is also accompanied by the accommodation of their schemas, but also by changes in the meanings of the instrument, resulting from the association of the artifact with new schemas' (our translation). Drawing once again from the same example, the scheme of 'striking an object' appears to be potentially associated with the wrench because of its properties, notably the fact that it has a mass at the end of a handle.

Thus, instrumentation also includes, in addition to the genesis of schemas, the dynamic processes of accommodation and assimilation that we have addressed above.

The development of instruments: A summary

The development of instruments can – and must – represent a relevant view for apprehending the relationship between workers and artifacts in an original way. This would account more fairly for the role played by the worker in the dynamic processes involved than does the view centred on interactions with artifacts. Here, we consider the worker at an individual level, but also at a level that is inscribed socially and collectively, a subject that bears a history and experiences, but who is placed in situations that are constantly renewed – and that are, themselves, more or less similar to situations that are already known – and who uses artifacts with specific properties – whether these properties are particularly rigid or not – through what he or she knows or thinks he or she knows, he or she can do with those artifacts (uses and potential uses) as a part of a finalized activity.

In favour of designing artifacts in an instrumental perspective

Instrumental genesis, described above, appears as something that is not only necessary but also unavoidable. Indeed, it could be viewed as a continuation of design, as the extension, within activity – taking into account its temporal features and its complexity – of classical design processes; the worker is then viewed as a 'designer in use'. Suggesting an encounter between these two design processes – design *for* use and design *in* use (Folcher, 2003) – makes it possible to open up some original and relevant prospects for designing artifacts with an instrumental approach. This will be the second orientation of the argument described in this chapter: the development of instruments is not just a process that is expected and observed, but it is also a goal and something that can be accompanied as part of a design project.

One explanation for the continuation of design in use lies with the fact that artifacts may be badly designed (Henderson, 1991; Thomas and Kellogg, 1989). Many of these design defects might, for example, be tied to the fact that designers have a relatively weak model of the worker.

The second explanation is that anticipation is obviously limited, because of the diversity of the workers' characteristics and the variability of situations. The goal of design then becomes anticipating more precisely, or offering a greater level of flexibility.

Some practical suggestions can be made to improve the design of artifacts:

- Providing workers with some room for manoeuvre by anticipating 'spaces of future possible activities' (Daniellou, 2004)
- Providing workers with artifacts that cannot be modified – or that can be modified, within the limitations set by designers, or alternately, that can be transformed following new perspectives from a functional point of view (Henderson and Kyng, 1991)
- Providing users with artifacts the design of which can be finished, based on boundaries that are set in design (Vicente, 1999)
- Readily providing artifacts that workers will, through their use, question and make evolve, either alone or in a collaboration with the stakeholders of design (Bannon and Bodker, 1991; Bodker, 1991)

The two explanations we have pointed out above rely on the fact that it seems impossible to anticipate all the possible uses of an artifact. However, the problem may not lie here.

Indeed, in the instrumental approach, design in use appears to be an intrinsic feature of the constitution of instruments. It is a process of design, but a process that this time is carried out by a worker involved in an activity. The processes of instrumental genesis do not reflect failures of design, but a stage that is necessary for taking ownership of artifacts, and for the development of these artifacts (Bannon and Bodker, 1991). This idea of incorporation can be viewed as a powerful driving force for the design of artifacts. The proposal of an instrumental approach is to open up a process that comprises loops, in order to start up a motion to integrate 'design in use' in the evolution of artifacts. This proposal of a loop can therefore be distinguished from a classical process model, where design and use are separated in time, and where the use stage is intended to be only the implementation of the artifact. On the contrary, what we have here is a reinclusion of the processes of instrumental genesis within the 'overall cycle of the design of an artifact' (Rabardel, 1995, p. 164, our translation). Thus, design appears to be a distributed process: professional designers and worker-designers both contribute to the process, taking into account their skills and their roles (Béguin, 2003; Bourmaud, 2006; Rabardel, 2001).

The model thus suggests that it can ground itself in the knowledge of the development of instruments to provide input for the design process – particularly in the form of an iterative design process. The model is entirely geared toward development. The final part of this chapter will provide an illustration of the model.

From instrumental genesis to (re)designing artifacts

The work we conducted and report was carried out within a television company in charge of broadcasting radio and television stations at the national and international levels. It aimed to design a new system to supervise networks. Supervision consists in ensuring the quality of broadcast networks and continuity in the service sold to the customers, notably through the following tasks: detecting breakdowns and anticipating incidents, diagnosing breakdowns, resuming service through remote actions, starting up maintenance operations, etc. A large number of events take place on these networks, leading to the setting off of alarms (about 1500 alarms set off over the course of a 24-hour period of work) in the supervision tool (which we will call SUPERVIS). The workers – or supervisors – are therefore tasked with understanding and solving the problems posed by these events and reported by the alarms, and their actions are carried out remotely through SUPERVIS.

An instrumental genesis

The supervisors must carry out multiple, varied tasks, but it is mostly the activity related to managing the alarms that will be of interest to us in this chapter. The process of dealing with an alarm can be divided into five stages:

1. The alarm is displayed in SUPERVIS's alarm manager window.
2. The detection of the alarm, i.e. the moment when the alarm *starts to exist* for the worker.

3. The interpretation of the alarm.
4. The processing of the alarm, which can itself be divided into several stages – taking into account the alarm, diagnosis, remote action and resolution.
5. The end of the alarm.

Instrumentation
We will describe here an example of a process of instrumental genesis in the fourth stage, specifically the stage of 'alarm resolution' (Bourmaud and Rétaux, 2002). Resolution amounts to changing the colour of the alarm line in SUPERVIS's alarm manager window. The colour shifts from red or magenta (these two colours indicate different levels of severity for alarms) to a more neutral colour, beige. An analysis of the activity of 12 supervisors provided evidence for three kinds of schema involved in resolving an alarm:

• Schema type A: The alarm is resolved immediately following its appearance. This is generally the case with false alarms, known alarms and untimely alarms.
• Schema type B: The alarm is resolved after one, several or all of the stages involved in alarm processing.
• Schema type C: The alarm is not resolved, although one, some or all of the stages involved in processing have occurred.

Schema types A and B are present in the activity of all 12 supervisors, whereas schema type C was only present in 5 supervisors. One can therefore define two groups of workers (see Table 11.1): workers exhibiting schema types A and B (group 1) and workers exhibiting all three schema types (group 2). We therefore conducted interviews to analyze more precisely the process of alarm resolution.

One of the seven supervisors in group 1 said, 'When I am done with processing an alarm, I have to resolve it ... that's the rule'. Another claimed, 'If I don't need to deal with this alarm because I have processed it, then I'll resolve it so I don't need to see it anymore ... so that it's not red anymore ... and most of all, to see the other alarms come through'. A final worker states, 'I need to have a clean screen to see the new alarms coming in'. In fact, it seems that schema types A and B account for (1) the instructions related

Table 11.1 Composition of the Groups of Workers, Based on the Schema Types Present in Their Activity

	Group 1	Group 2
Schema types present in the activity of the workers	A + B	A + B + C = all
Number of workers	7	5

to resolving the alarms ('resolve alarms once all the required actions have been performed') and (2) the ease and speed of detection of incoming alarms, because of the contrast between the old alarms (which have been 'beiged out') and new alarms (which are coloured red or magenta).

The interviews carried out with the five supervisors in group 2 provide some additional, most interesting information about the process of instrumentation. One of the supervisors claimed he 'preferred not to resolve any alarms to keep them around … to always have them in sight'.

Before we go on, however, what are the consequences of failing to resolve some of the alarms on the 'alarm manager' window? Two points should be made here:

1. Unresolved alarms continue to be displayed at the bottom of the screen. Because the alarms are displayed in chronological order, the more recent alarms are displayed in the bottom-most part of the screen.
2. When the alarms are resolved, they are displayed above the unresolved alarms. Upon arrival, new alarms are therefore displayed on the screen just above the 'conserved' alarms. There are typically few alarms of this kind (between three and eight alarms for the entire duration of the workday, for each supervisor in group 2). The activity of the supervisors does not seem to be hindered by this. There is no confusion between the unresolved alarms, whether they have been processed or not. The supervisors act only on new alarms. The verbal utterances of the workers in group 2 suggest that what is important to them is to be able to tell alarms apart. Some of the alarms are resolved, and those that are unresolved remain at the bottom of the screen.

Therefore, the C type schema leads to keeping some alarms grouped in a specific location. The alarms that are viewed as particularly important (or critical) are therefore highlighted, which helps to supervise and follow them. Some other elements seem to confirm and extend this analysis.

First, in-depth interviews and the analysis of work diaries – the documents where significant events having occurred in the course of work are recorded – show that there is a correlation between the alarms that are 'set aside' and those that are recorded in these diaries. The latter alarms, generally, are alarms that have proven to be impossible to resolve via remote actions, and in some cases led to setting off a maintenance operation or specific processing. Approximately 9 out of 10 'conserved' alarms were also recorded in the diaries. Furthermore, because the supervisors worked continuous shifts, there were handovers and, consequently, some exchanges of instructions. Many instructions were exchanged between the two workers, who relayed to one another. Thus, as shown in Table 11.2: (1) when the worker finishing his shift belonged to group 1, he relied

Table 11.2 Alarms Processed during the Handover
of Instructions, Depending on the Instrument Used[a]

	Alarms processed during the handover of instructions, depending on the instrument used[b]	
	Group 1	Group 2
Work diary	85%	40%
SUPERVIS	0%	40%
Both	15%	15%
Neither	0%	5%

[a] Per group, in per cent of alarms.
[b] In per cent of alarms.

mostly on the work diary when handing over instructions (for 85 per cent of all alarms processed), or alternately, both on the diary and on the alarms resolved in the alarm manager (for 15 per cent of alarms); (2) when the worker finishing his shift belonged to group 2, he relied on the diary (for 40 per cent of alarms), on the alarm manager screen (for 40 per cent of all alarms) or on both these sources (for 15 per cent of all alarms).

Finally, we observed one last strong correlation. The C type schema appeared mainly in the activity of the supervisors with the greatest expertise. We defined two categories depending on their level of expertise, based on the cross-examination of two parameters: seniority in the position and peer recognition. Five of the six workers belonging to the category of expert workers were also placed in group 2, and conversely, all the workers belonging to the category of nonexpert workers belonged to group 1.

Instrumentalization

SUPERVIS is a system that can only be modified very little. Possibilities for configuration and customization provided to the workers are minimal. However, supervisors can request adaptations and alterations to other workers – the configurers – who are in charge of carrying out the alterations in SUPERVIS, related to evolutions in the organization of the company, customer requests, the technology used in the equipment, etc. These requests are typically dealt with by the configurers themselves, who integrate them within the tool.

The design process

The supervisors' activity of design in use seems to be mainly geared toward the scheme component of the instrument (i.e. instrumentation) rather than toward its artifact component (i.e. instrumentalization). However, the use of SUPERVIS, which aims mostly to highlight critical alarms, appeared to be very relevant in the design team for NEW-SUPERVIS – the artifact that subsequently replaced SUPERVIS, and contributed to the specifications of this new artifact.

Recovering design in use
First, it was decided to salvage the product of this design in use, and to turn it into an intrinsic function of NEW-SUPERVIS. Hence, the design team agreed on the specifications of a window dedicated to the alarms that the supervisors deemed were 'special'. The group chose to call this window the *alarm list*. Supervisors may place an alarm in the list (or remove it from the list), thus benefitting from a function that is new and powerful to support their activity.

Supporting design in use
The ability of supervisors to tailor SUPERVIS to their own activity surprised the entire design team, including the designers and the workers themselves. A required space that had been left to the supervisors to configure and personalize NEW-SUPERVIS was viewed by all to be an interesting and useful feature. The team therefore agreed on the principle of making NEW-SUPERVIS more easily adaptable and customizable. This perspective follows in part the proposal of Henderson and Kyng (1991), presented above: to provide workers with artifacts that can be adapted and modified, as instrumental proposals that they may or may not implement. The goal is to propose 'plastic' artifacts allowing the worker to organize his own instrumental geneses. Each worker might then be able to construct his own instrument.

Conclusion

The goal of this chapter was twofold. The first goal was to present a particularly elaborate approach to the worker-artifact relationship, using the concept of instrument. The instrument is not only an entity that is external to the worker, to which he must confront himself from the points of view of interaction and time (related to a stage of appropriation). The instrument also points to an element that is internal to the worker, as an entity that is part of the worker through his activity. The instrument then emerges as a resource that is developed by workers themselves, to be mobilized in their environment, for a specific goal. The relationship between the instrument and the worker seems to act both ways, although it is initially guided by the worker. Each of the two contributes to transforming the other, transforming oneself in the process.

The second goal related to the very issue of designing artifacts – where designing artifacts in an instrumental approach may become a stake for ergonomic action. This approach of design places the resources developed by workers as prior constructions, to be put to the service of an efficient, high-quality (re)design of artifacts.

Acknowledgements

The author thanks Françoise Decortis for her valuable advice.

References

Bannon, L., and Bodker, S. (1991). Beyond the interface: encountering artifacts in use. In J. Carroll (Ed.), *Designing interaction: psychology at the human computer interface*. Cambridge: Cambridge University Press.

Béguin, P. (2003). Design as a mutual learning process between users and designers. *Interacting with Computers*, 15(5), 709–730.

Bodker, S. (1991). *Through the interface: a human activity approach to user interface design*. Mahwah, NJ: Lawrence Erlbaum Associates.

Bourmaud, G. (2006). *Les systèmes d'instruments: methodes d'analyse et perspectives de conception*. Unpublished doctoral dissertation, Université Paris 8.

Bourmaud, G., and Rétaux, X. (2002, November). Rapports entre conception dans l'usage et conception institutionnelle. Presented at 14th French-Speaking Conference on Human–Computer Interaction, Poitiers, France.

Cole, M. (1996). *Cultural psychology: once and future discipline?* Cambridge, MA: Harvard University Press.

Daniellou, F. (2004). L'ergonomie dans la conduite de projets de conception de systèmes de travail. In P. Falzon (Ed.), *Ergonomie* (pp. 358–373). Paris: PUF.

Daniellou, F., and Rabardel, P. (2005). Activity-oriented approaches to ergonomics: some traditions and communities. *Theoretical Issues in Ergonomics Science*, 6(5), 353–357.

Faverge, J. M. (1977). *Analyse de la sécurité du travail en termes de facteurs potentiels d'accidents*. Working Paper of the Industrial Psychology Laboratory. Brussels: Université Libre de Bruxelles.

Folcher, V. (2003). Appropriating artifacts as instruments: when design-for-use meets design-in-use. *Interacting with Computers: The Interdisciplinary Journal of Human Computer Interaction*, 15(5), 648–663.

Folcher, V., and Rabardel, P. (2004). Hommes, artefacts, activités: perspective instrumentale. In P. Falzon (Ed.), *Ergonomie* (pp. 251–268). Paris: PUF.

Gibson, J. J. (1979). *The ecological approach to visual perception*. Boston: Houghton Mifflin.

Henderson, A. (1991). A development perspective on interface, design and theory, in designing interaction. In J. Carroll (Ed.), *Psychology at the human computer interface* (pp. 254–268). Cambridge: Cambridge University Press.

Henderson, H., and Kyng, M. (1991). There's no place like home: continuing design in use. In J. Greenbaum and M. Kyng (Eds.), *Design at work: cooperative design of computer systems* (pp. 219–240). Hillsdale, NJ: Laurence Erlbaum Associates.

Kaptelinin, V., and Kuutti K. (1999). Cognitive tools reconsidered. From augmentation to mediation. In J. P. Marsh, B. Gorayska, and J. L. Mey (Eds.), *Human interfaces: questions of method and practice in cognitive technology.* Amsterdam: Elsevier Science B.V.

Kaptelinin, V., and Nardi, B. A. (2006). *Acting with technology: activity theory and interaction design.* Cambridge, MA: MIT Press.

Leontiev, A. N. (1978). *Activity, consciousness, and personality.* Englewood Cliffs, NJ: Prentice-Hall.

Norman, D. A. (1988). *The psychology of everyday things.* New York: Basic Books.

Norman, D. A. (1991). Cognitive artifacts. In J. M. Caroll (Ed.), *Designing interaction: psychology of human computer interface* (pp. 17–38). Cambridge: Cambridge University Press.

Piaget, J. (1952). *The origins of intelligence in children.* New York: W. W. Norton.

Rabardel, P. (1995). *Les hommes et les technologies, approche cognitive des instruments contemporains.* Paris: Armand Colin. An English version may be retrieved from http://ergoserv.psy.univ-paris8.fr/Site/default.asp?Act_group=1.

Rabardel, P. (2001). Instrument mediated activity in situations. In A. Blandford, J. Vanderdonckt, and P. Gray (Eds.), *People and computers XV – interactions without frontiers* (pp. 17–30). Berlin: Springer-Verlag.

Rabardel, P., and Bourmaud, G. (2003). From computer to instrument system: a developmental perspective. *Interacting with Computers,* 15(5), 665–691.

Rabardel, P., and Waern, Y. (2003). From artefact to instrument. *Interacting with Computers: The Interdisciplinary Journal of Human Computer Interaction,* 15(5), 642–645.

Suchman, L. (1987). *Plans and situated actions: the problem of human-machine interaction.* Cambridge: Cambridge University Press.

Thomas, J., and Kellogg, W. (1989). Minimizing ecological gaps in user interface design. *IEEE Software,* 78–86.

Vicente, K. J. (1999). *Cognitive work analysis: toward safe productive and healthy computer-based works.* Mahwah, NJ: Lawrence Erlbaum Associates.

Vygotsky, L. S. (1931–1978). *Mind in society: the development of higher psychological processes.* Cambridge, MA: Harvard University Press.

Wertsch, J. V. (1998). *Mind as action.* New York: Oxford University Press.

chapter twelve

Prevention of MSDs and the development of empowerment

Fabien Coutarel and Johann Petit

Contents

The prevention of work-related musculoskeletal disorders[*] (MSDs) is a classical object for ergonomics. It is also, perhaps, one of the most internationalized topics in the field of occupational health and safety.

Since the 1990s, the world of work has undergone some major evolutions. The various forms of intensification of work, widely described in the literature, have led to an 'explosion' of MSDs. What may be termed 'industrialization of services' or 'tertiarization of the industrial world' refers to a multiplication of constraints in all fields of human activity. The proportion of French workers performing their job under constraints of work rate is increasing (Arnaudo et al., 2010). In addition to the

---

[*] In the scientific literature, MSDs refer mostly to pathologies of the upper limbs. Because of the proximity between aetiological processes and the means of action on the work environment, pains in the lower back and in the neck tend to be included in what ergonomics refers to as MSDs.

usual constraints of the industrial world, one can also find constraints from the service world (quality, relations with customers and providers, audits, tailored services with low delays, etc.). In the world of services, the individualized evaluation of work imposes growing quantitative requirements that mistreat the service relation itself – and consequently mistreat its agents. MSDs today represent more than 80 per cent of occupational diseases in terms of compensations received.

In the 1980s and 1990s in France, as in other parts of the world, the dominant approaches to the prevention of MSDs focused on the biomechanics of movement, underlying pathogenic conditions in the solicitation of the body at work in terms of intensity of posture, duration, frequency and vibration.

Although these approaches are essential, they have proven to be incapable of ensuring prevention on their own, for two crucial reasons:

- On the one hand, because these approaches focus on the pathogenic mobilization of the body, the possibilities they afford for transforming work situations have been restricted to designing the 'proximal means' of work (workstations and work tools) and to training workers to achieve correct gestures and postures.
- On the other hand, these approaches have underestimated the multifactorial character of the pathology. Relations of the pathology with the other dimensions of mobilization at work have, since then, been noted by both practitioners and researchers (Bongers et al., 2006; National Research Council, 2001; Kausto et al., 2010; Krause et al., 2010; Van Rijn et al., 2010). Work in epidemiology has also led to a complexification of the models of MSD aetiology. These models gradually integrated other factors, which were termed psychosocial and organizational (autonomy, collective support, organization, workload, etc.). Today, the literature acknowledges the importance of driving forces for transformation, located at the level of the organization and of the design of work systems (including the design of the design process itself).

Activity ergonomics and the prevention of MSDs: A developmental approach

On one occasion where he described activity theories, their development and their evolution, Wisner noted that 'the broadening of the scope of action is a typical and fundamental feature of human development' (1997, p. 50, our translation). Further developments later came to enrich and specify this developmental dimension (Coutarel and Daniellou, 2011): whereas activity was initially centred on the completion of a job, it gradually became a 'slice of life' where the individual also stakes his or her own subjectivity in the relation to work (Van Belleghem et al., this volume).

Activity ergonomics (Daniellou, 2005; Daniellou and Rabardel, 2005) leads us to defend an original developmental position in the international context of work on MSDs, which is, to this day, largely dominated by a type of ergonomics that is termed physical.

The general argument that we will make here can be summarized in the following way: the development of professional activities in and through an intervention constitutes the main means for the prevention of MSDs within the scope of ergonomic action.

In this view, the design of workstations, project management, etc. are no longer goals for the ergonomic intervention, but means to produce development. In this sense, this is an inverse perspective to the work that considers the participation of workers as a means to achieve other goals (Kuorinka, 1997; Wilson and Haines, 1997).

Obviously, we are not the first authors to have placed out work within this developmental approach to the prevention of MSDs. Depending on the author, this approach is made more or less explicit. Several works in the clinical approach of activity, in clinical medicine, in ergology and in ergonomics have placed this issue of preventing adverse effects on the health of workers in terms of operational leeway (Coutarel, 2004; Clot and Fernandez, 2005), hyposocialization of gestures (Simonet, 2011), hindrances (Sznelwar et al., 2006), chronic inhibition of subjectivity (Davezies, 2011) and drama of self-use (Schwartz, 2010).

In the tradition of activity ergonomics, this orientation has led to programs aiming to train the stakeholders of the intervention to ergonomic work analysis in order to make them capable of updating, whenever necessary, their knowledge of the system and of managing future projects (Daniellou, 2004; Daniellou and Martin, 2007; Falzon and Mollo, 2009; Garrigou et al., 1995). Furthermore, although their work does not deal specifically with MSDs, our approach also converges with the theoretical proposals of Rabardel and Béguin (2005) and Nathanael and Marmaras (2008), whose work on the processes of designing and introducing new technologies precisely demonstrates the issues behind the joint development of professional activities.

Turning the development of professional activities into the main driving force for preventing MSDs implies some consequences regarding how prevention should be viewed; the meaning of work to those who carry it out is at the heart of this approach. Turning prevention into a separate issue, one that is set aside from work (Coutarel, 2011), significantly reduces not only the resources that can be mobilized for the transformation of work, but also the relevance of the measures that are implemented by the stakeholders themselves. There is no such thing as prevention of MSDs without an understanding of activity from the point of view of those who carry it out. Issues related to performance (quality, customer satisfaction, mutual aid between workers) constitute some of the central dimensions of this activity. This position firmly distinguishes the developmental approach from the hygienist approach, which focuses prevention on reducing the constraints of exposure. Instead, the goal of the developmental approach is to develop the resources of agents and organizations for coping with the everyday challenges of work, to foster the possibility of achieving pro- duction goals in favourable conditions. There is no other way to explain why, for example, a cashier speeds up of her own accord when she sees a large queue waiting, nor is there any way to propose relevant solutions for prevention. If the cashier must choose between satisfying the customer and limiting the repetitiveness of her gestures, she will choose quality of service, because that is where the meaning of her work lies.

We will theorize this developmental approach around two main con- cepts: leeway and empowerment. In this model, leeway is situational. It constitutes a space for the regulation of activity, resulting from an encoun- ter between a system of constraints (external leeway), on the one hand, and an individual or a collective, on the other hand, in a given work situ- ation. Empowerment refers to an active relationship between the individ- ual and his or her environment. This concept is different from leeway in two ways: temporality and the perimeter of the situations involved.

A situational approach to MSDs: Between external leeway and internal leeway

In the tradition of activity ergonomics, leeway constitutes a space for regulating activity. This space is the result of an encounter between the characteristics of the professional environment and those of the worker(s) involved (Coutarel, 2004; Durand et al., 2008). Leeway reflects the active relationship between the individual and his or her task. The possibility of being involved in one's work constitutes both a means for personal devel- opment and a condition of performance.

Within the field of clinical studies of activity, whose proximity with activity ergonomics is well known, the concept of empowerment echoes that of leeway. This connection is so strong that some ergonomists tend to switch from one to the other with no particular caution. We will propose here a model allowing us to describe in greater detail this concept of lee- way, derived from activity ergonomics, specifically in terms of its connec- tion with empowerment (Clot, 2008).

This leeway is situational. As shown in Figure 12.1, it depends on the specific features of the work situation of interest, and it can be characterized:

- On the one hand, in terms of internal leeway, which is perceived and constructed by the individual depending on his characteristics at the time
- On the other hand, in terms of the external leeway, which is constructed by the sociotechnical and organizational environment

Internal leeway and external leeway are not independent from one another. For example, a new skill might introduce a new position, leading the sociotechnical system to change the rules governing the work of the worker involved. Conversely, evolutions in the organization may lead to new possibilities of mobilizing the know-how that has been newly acquired.

According to this model, an ergonomic intervention may contribute to the development of situational leeway by acting:

- On internal leeway (training, mobilization, perception of other possible futures, perceived state in any given instant, previous pains) where the subjectivity of agents, which we will define here as the ability to be affected by the experience of work (Coutarel and Daniellou, 2011), constitutes a target for the intervention.
- On external leeway, which is determined in general by the ability of a work organization and its related tools to manage variability within work. There are numerous possibilities for acting on the external leeway: supporting mutual aid and sharing about work, supporting

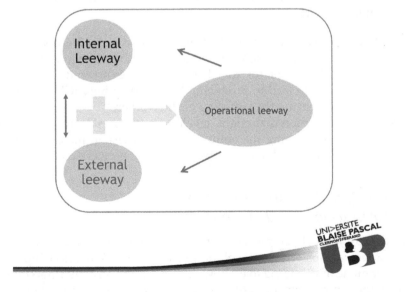

Figure 12.1 Situational leeway.

the reelaboration of rules (Caroly, 2010), designing the work process, the maintenance of installations and equipment, the management of human resources, forms of management, customer-provider relations, project management, etc.

The situational leeway thus depends on a circumstantial encounter between the various determinants of a work situation that make up the internal and external leeway. To develop leeway implies developing the resources that are available in the situation – and therefore the possibilities for the workers involved to respond to the requirements of work (including those requirements that they set for themselves) in physical, organizational and social conditions that allow taking into account the various modes of their mobilization (subjective, cognitive, physiological and biomechanical). Thus viewed, the development of situated leeway transforms work, leading, as a consequence, to a reduction in the risk of exposure to MSD-inducing factors.

Hence, there are many stakeholders that are directly concerned with leeway within the company: workers, middle management, supervisors of the various departments and company management. Often, the lack of leeway for some workers results from the difficulties of other stakeholders – for example, those in charge of managing the organization and preserving that very leeway. Thus, inevitably, the prevention of MSDs involves various actors in the work environment.

There are two consequences for ergonomic interventions that should be highlighted here:

- Any intervention that would develop the internal leeway of agents without taking into account the tolerance that is required from the work environment for this leeway to be expressed (i.e. external leeway) would lead to producing hindered activity (Clot, 1999), and thus to increasing the suffering of the workers.
- Any intervention that might develop the external leeway of workers without developing the ability of the workers to make use of this leeway would lead to little effective impact on work conditions.

A developmental approach to MSDs: Between leeway and empowerment

The developmental perspective may aim to allow the sustained construction of leeway, fostering an evolution of the worker's relationship to the environment, and leading to empowerment. Empowerment thus suggests a sustained improvement in the scope of action. The ergonomic intervention takes on a developmental perspective, and is then viewed as a means to foster empowerment. Its scope goes beyond that of the intervention

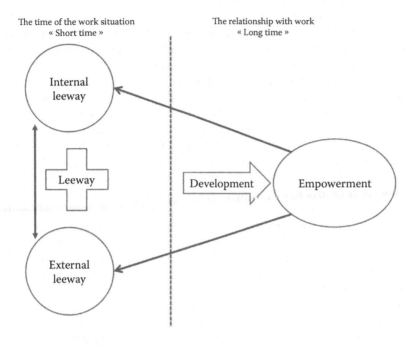

Figure 12.2 Leeway and empowerment.

and the construction of the 'pretext situation' that is to be designed (Figure 12.2).

Therefore, situational leeway is included in a relationship with the work environment – a relationship that it helps to build. Here, empowerment takes place within a history, whose time frame and scope go beyond that of the work situation. Thus, a situation with no leeway does not necessarily entail a reduction in empowerment. A situation with increased leeway may contribute to the development of empowerment, but this development also depends on many other work situations and life situations within his professional environment.

Conversely, the development of empowerment leads to considering that change is necessary – including for situations that had previously not been taken into account. In this sense, empowerment fosters the creation of new leeway. It is the perception of other possibilities that turns the situation into an unbearable one (Sartre, 1943).

By radicalizing this position, the intervention project becomes a pretext to introduce sustainably into the system, new relations for negotiating the work environment that are more supportive of the expression of subjectivity in work. In this sense, the intervention is a political tool (Hubault and Bourgeois, 2004). Empowering workers implies durably strengthening their ability to influence their work situations and renor-malize the professional environment (Schwartz, 2010). Switching from the goal of developing situational leeway to that of developing empower-ment introduces within the ergonomic intervention, on the one hand, the durable aspects of the relation with the environment, and on the other hand, the stake of transforming the human relations at work. The goal is to develop, within the intervention, the ability of agents to transform the work environment, for the widest possible range of reasons. This implies knowledge of the company and its workings, its agents, the constraints, the stakes involved, the work of others, etc. This also implies the creation of meaning, for which the collective constitutes a key resource (Clot, 1999), leading the intervention to focus more on the stakes of collective work.

Let us take the example of postural constraints in the upper limb. The fundamental issue is not to limit the range of situations where the worker will work with arms raised. The fundamental questions are: Why is the organization incapable of managing the fatigue experienced by the workers and associated with a job that is carried out with arms raised in the air? What is the place given to the experience of workers in managing everyday work? What are the organizational tools that do not work, for fatigue to be no longer expressed, managed or anticipated in design? The development of empowerment serves the prevention of MSDs if it contrib-utes to increasing the power of experience in managing work, managing it in everyday settings and managing its projects. We argue that the tools considered here are mostly of a cognitive nature; the issue is to develop the ability of the organization to apprehend work in all of its complex-ity. This organizational ability is a truly political ambition with respect to work, which must be supported by management tools, and for which these tools cannot substitute.

From this point of view, MSDs are only one of many possible symp-toms of an organizational malfunction; work organizations and those who manage them are in the forefront when the issue is to interrogate the leeway of the agents. From this point of view, MSDs reflect an inability of the agents of the organization to manage the various stakes of work in various moments of life at work. This inability manifests itself in modes of organizational management that do not take into account the work activity and, in particular, its subjective aspects (Coutarel and Petit, 2009). The intervention is an opportunity to contribute to the development of the activity of various stakeholders: workers, designers, decision makers and staff representatives (Barcellini et al., this volume). From this point of view, it is beyond doubt that the topic of MSDs loses much of its speci-ficity, in relation to other topics of occupational health and safety, such as psychosocial hazards (Coutarel, 2011; Petit et al., 2011). Here, therefore, the ergonomic intervention is an opportunity – a pretext? – to the devel-opment of professional activities (and hence of individuals) destined to foster the sustained development of the ability of organizations to manage work and its unforeseen events. This is an a maxima vision that we will propose for the prevention of MSDs and the role of ergonomics. The goal is not just to reduce constraints or exposure to risk factors, as is most often suggested in the international literature. It is a change in paradigm where prevention rests first on the development of agents, through the development of their professional activities.

Design situations are, no doubt, privileged moments to develop empowerment in agents. The distinctive status of the design project affords easier opportunities to implement innovative forms of management, whose relevance must then be demonstrated, along with their implications for project management, where the wealth of the organization lies not so much in its concrete form – whatever that form may be – but rather in its ability to evolve in a context of constant change.[*]

Materializing this ambition of developing empowerment in the ergonomic intervention probably implies forms of ergonomic intervention that are related more to 'having the organization do the job' and 'supporting the change' rather than 'doing it on one's own', which would imply that the agents of the organization would be capable of deciding a change in organizational management beyond the scope of a project. This type of expertise in ergonomists is grounded in the idea that the project constitutes an opportunity to try out other ways of thinking about work, conducting change and managing everyday work. It suggests a strong commitment of the company stakeholders in the intervention. Eliciting the scope of the intervention (i.e. the development of situational leeway/ development of empowerment) must make it possible to justify the added requirements requested from the agents of the professional environment as part of the intervention.

The general postulate can be specified through the following model:

1. The development of professional activities in and through the intervention constitutes the main driving force of the prevention of MSDs for ergonomic interventions.
2. The ergonomic intervention contributes to the development of professional activities through the deployment of leeway in work situations. This leeway is the product of an encounter between external leeway (the plasticity of the work system) and internal leeway (abilities of individuals). It aims to support, in satisfactory conditions, the completion of performance goals by the workers.

---

[*] We have in the past used the term *enabling organization* (Coutarel and Petit, 2009), echoing the concept of enabling work (Falzon and Mollo, 2009).

3. The way in which the ergonomic intervention is conducted may produce effects on the agents involved that go beyond the confines of the initial situation. The sustainable alteration of relations with the professional environment, fostering a greater ability of subjectivity to influence the course of work, contributes to the development of empowerment.

Case study: An ergonomic intervention in the food processing industry

The case of an intervention to design a warehouse for cutting fattened ducks (Coutarel, 2004) is a good illustration of our proposals, since we were able to reconstruct the history of the company involved over the course of approximately 10 years.

The company was confronted both with a demand for increased production and with increasingly numerous complaints related to MSDs. Hence, it wished to include ergonomists in a project aiming to design a new production warehouse, to both respond to its ambition of increasing its production capacity and prevent MSDs. A project management scheme, involving various stakeholders (workers, middle and upper management, contractors and institutional stakeholders) was implemented. The duration of the project was 2½ years, starting with the early analyses of work situations in existing situations, up to the implementation of the new systems. The design of the new technical systems and new modes of work organization (both for the management of everyday work and for the management of the project) led to obtaining positive results for the health of the workers (assessed through a clinical examination by the occupational physician and through questionnaires) and for system performance (evaluated using the company's performance indicators). In terms of performance, results exceeded the initial goals and divided the time before return on investment (ROI) by 3. Faced with this result, the company decided – in spite of our warnings – to buy the same production line from the supplier, to implement it in its other production site. On this second site, the expected goals in terms of performance and health were never achieved.

A few months later, the company encountered some financial difficulties, leading it to impose new standards for production and work organization. These standards questioned the design trade-offs that had been made previously as a part of the ergonomic intervention. The situation of the workers was disrupted, leading to a new surge in MSDs. In the end, the experience of the workers in the course of the ergonomic intervention made it more difficult to return to more traditional standards of operation, with respect to the usual practices in the industry (production rates, product per worker per hour ratio, etc.). Returning to the company some years later allowed us to notice a new reversal in the situation: the agents who had been a part of the initial ergonomic intervention, who were all still present in the workforce, had been able to take advantage of or create modes of work that were close to the trade-offs that had been made originally, while also developing projects following the principles of project management that had been tested in the initial intervention (Dugué et al., 2010).

Although we would not want to draw some overly generalized principles from this case, we are nevertheless led to believe that the process of involving these agents:

1. Allowed the development of leeway for agents in the work situations that had been covered in the initial projects. Results proved to be positive in terms of health and performance. The failure to achieve the same results in the other production site, in spite of implementing the same tools and procedures, highlights the importance of the intervention process, compared with the technical and organizational specifications that are produced in the design project.
2. Fostered sustained transformations in the way stakeholders envisioned their everyday work and the conduct of organizational change, thus contributing to the development of their empowerment. This allowed the agents to once again mobilize their resources, in new work situations, and in new projects, where new situational leeway was created. Leeway is therefore both a condition and a consequence of empowerment.
3. This development of empowerment affects individuals. This may involve uncomfortable or critical stages.

The external validity of the model centered on leeway and empowerment

The external validity of the model can be apprehended through three key criteria: the evolution of aetiological models, the changing features of current work situations and the context of the knowledge economy.

The evolution of knowledge regarding the aetiology of MSDs currently emphasizes the relevance of a global and systemic approach to work, where the forms of mobilization of agents in and through the intervention have a decisive impact on the results achieved, in terms of both prevention and performance of the system considered (Petit and Coutarel, this volume).

Indeed, it seems that the international consensus is set on the following two points:

- The complex and multifactorial nature of the processes leading to pain, justifying a multidimensional approach of the prevention intervention. This intervention should cover the entirety of the work situation and the company, in order to allow envisioning a simultaneous, coordinated action on constraints and their determinants (Dempsey, 2006; Kennedy et al., 2010; Laing et al., 2007).
- The relation to work, justifying interventions in the real-world work environment (Da Costa and Vieira, 2009; Ijzelenberg et al., 2004; Ostergren et al., 2005; Punnett and Wegman, 2004; Wells, 2009).

The constant change of work situations (flexibility, reorganizations, low job security, modes of production management, etc.) (Dugué, 2006) is such that the benefits of technical interventions aiming to reduce the physical constraints of the work situation are increasingly short-lived. This new feature of contemporary work obviously leads to ergonomics readjusting its scope out of necessity. What is the return on investment of investing in ergonomics, if its contribution is limited to designing work situations whose characteristics evolve increasingly quickly? Already, Kuorinka (1998) emphasized the need for 'quick and flexible strategies' as a direct consequence of the constant changes of production environments, to improve the prevention of MSDs.

The context of the economy of knowledge described by Foray (2009) might, today, foster a renewed interest for integrating human factors in the management of organizations, where the worker – including his subjectivity – is apprehended as the primary resource (Lièvre and Coutarel, 2013). Indeed, the system used for the economic valuation of work influences the way in which ergonomic recommendations and prescriptions are received, as well as their very nature (Hubault and Bourgeois, 2004).

Current and future epistemological challenges

It remains quite difficult, even today, to ensure that the results obtained through a developmental approach to the prevention of MSDs are recognized by the international community. The main obstacle to this seems, to us, to be of an epistemological nature (Coutarel et al., 2005): in the face of an epistemology where the dominant view lies with the control of factors, complex interventions (Champagne et al., 2009) that rely on an epistemology of complexity find it difficult to exist. As evidence, one need only examine the work that is mentioned in reviews of the literature on the subject, and notably the selection criteria that are used: randomization, control group and statistical validity (Driessen et al., 2010, 2011; Rivilis et al., 2008; Roquelaure, 2008; Tuncel et al., 2008).

A large-scale epistemological work is now required, so that the literature reviews, which serve as a means to define the state of knowledge regarding the effectiveness of interventions, might include the results of our work: to develop a model for the assessment of complex interventions, grounded in an epistemology of complexity, which might specify the operational conditions required for the production of knowledge based on complex cases. The lack of detail in descriptions of such interventions (Cole et al., 2003) and the diversity in the modes of description of these interventions are obstacles to generalization (Denis et al., 2008; Kristensen, 2006). Only upholding this level of epistemological strictness can lead to methodologies derived from fields other than epidemiology to take their place in systematic reviews of the literature, without being viewed as too weak (Neumann et al., 2010).

References

Arnaudo, B., Léonard, M., Sandret, N., Cavet, M., Coutrot, T., and Rivalin R. (2010). L'évolution des risques professionnels dans le secteur privé entre 1994 et 2010. *DARES Analyses*, 23.

Bongers, P. M., Ijmker, S., Van den Heuvel, S., et al. (2006). Epidemiology of work related neck and upper limb problems: psychosocial and personal risk factors (part I) and effective intervention from a bio behavioural perspective. *Journal of Occupational Rehabilitation*, 16, 279–302.

Caroly, S. (2010). L'activité collective et la réélaboration des règles: des enjeux pour la santé au travail. Habilitation thesis, Université Bordeaux 2, France.

Champagne, F., Contandriopoulos, A. P., Brousselle, A., Hartz, Z., and Denis, J. L. (2009). L'évaluation dans le domaine de la santé: concepts et méthodes. In A. Brousselle, F. Champagne, A. P. Contandriopoulos, and Z. Hartz (Eds.), *L'évaluation: concepts et méthodes*. Montréal: Presses de l'Université de Montréal.

Clot, Y. (1999). *La fonction psychologique du travail*. Paris: PUF.

Clot, Y. (2008). *Travail et pouvoir d'agir*. Paris: PUF.

Clot, Y., and Fernandez, G. (2005). Analyse psychologique du mouvement: apport à la compréhension des TMS. *Activités*, 2(2), 68–78.

Cole, D. C., Wells, R. P., Frazer, M. B., Kerr, M. S., Neumann, W. P., Laing, A. C., and the Ergonomic Intervention Evaluation Research Group. (2003). Methodological issues in evaluating workplace interventions to reduce work-related musculoskeletal disorders through mechanical exposure reduction. *Scandinavian Journal of Work, Environment and Health*, 29, 396–405.

Coutarel, F. (2004). La prévention des troubles musculo-squelettiques en conception: quelles marges de manœuvre pour le déploiement de l'activité? Unpublished doctoral dissertation, Laboratoire d'Ergonomie des Systèmes Complexes, Université Bordeaux 2, Bordeaux.

Coutarel, F. (2011). Des "TMS" aux "RPS", quand tout nous invite à parler "Travail". In F. Hubault (Ed.), *Risques psychosociaux: quelle réalité, quels enjeux pour le travail?* (pp. 99–119). Toulouse: Octarès.

Coutarel, F., and Daniellou, F. (2011). L'intervention ergonomique pour la prévention des troubles musculosquelettiques: quels statuts pour l'expérience et la subjectivité des travailleurs ? *Travail et Apprentissages*, 7, 62–80.

Coutarel, F., Daniellou, F., and Dugué, B. (2005). La prévention des troubles musculo-squelettiques: des enjeux épistémologiques. *Activités*, 3(2), 3–19.

Coutarel, F., and Petit, J. (2009). Le réseau social dans l'intervention ergonomique: enjeux pour la conception organisationnelle. *Revue Management and Avenir*, 27, 135–151.

Da Costa, B. R., and Vieira, E. R. (2009). Risk factors for work-related musculoskeletal disorders: a systematic review of recent longitudinal studies. *American Journal of Industrial Medicine*, 53, 285–323.

Daniellou, F. (2004). L'ergonomie dans la conduite de projets de conception de systèmes de travail. In P. Falzon (Ed.), *Ergonomie* (pp. 359–373). Paris: PUF.

Daniellou, F. (2005). The French-speaking ergonomist's approach to work activity; cross-influences of field intervention and conceptual models. *Theoretical Issues in Ergonomics Science*, 6(5), 409–427.

Daniellou, F., and Martin, C. (2007, March). *Quand l'ergonome fait travailler les autres, est-ce de l'ergonomie? Journées de Bordeaux sur la Pratique de l'Ergonomie*. Bordeaux, France.

Daniellou, F., and Rabardel, P. (2005). Activity-oriented approaches to ergonomics: some traditions and communities. *Theoretical Issues in Ergonomics Science*, 6(5), 353–357.

Davezies, P. (2011). Souffrance sociale, répression psychique et troubles musculosquelettiques. Presented at 3rd French-Speaking Congress on Musculo-Skeletal Disorders, Grenoble, France. Retrieved from http://halshs. archives-ouvertes.fr/halshs-00605360/.

Dempsey, P. G. (2006). Effectiveness of ergonomics interventions to prevent musculoskeletal disorders: beware of what you ask. *International Journal of Industrial Ergonomics*, 37, 169–173.

Denis, D., St-Vincent, M., Imbeau, D., Jette, C., and Nastasia, I. (2008). Intervention practices in musculoskeletal disorder prevention: a critical literature review. *Applied Ergonomics*, 39, 1–14.

Driessen, M., Proper, K. I., Anema, J. R., Knol, D. R., Bongers, P. M., and van der Beek, A. J. (2011). The effectiveness of participatory ergonomics to prevent low-back and neck pain – results of a cluster randomized controlled trials. *Scandinavian Journal of Work, Environment and Health*, 37(5), 383–393.

Driessen, M., Proper, K. I., van Tulder, M. W., Anema, J. R., Bongers, P. M., and van der Beek, A. J. (2010). The effectiveness of physical and organizational ergonomic interventions on low back pain and neck pain: a systematic review. *Occupational and Environmental Medicine*, 67, 277–285.

Dugué, B. (2006). La folie du changement. In L. Théry (Ed.), *Le travail intenable* (pp. 95–118). Paris: La Découverte.

Dugué, B., Chassaing, K., and Coutarel, F. (2010, September). Work-related musculoskeletal disorders prevention: assessment of an ergonomic intervention 6 years later. Presented at PREMUS Congress Proceedings, Angers, France.

Durand, M. J., Vézina, N., Baril, R., Loisel, P., Richard, M. C., and Ngomo, S. (2008). *Étude exploratoire sur la marge de manœuvre de travailleurs pendant et après un programme de retour progressif au travail: définition et relation(s) avec le retour en emploi.* Collection Études et Recherches, IRSST, Project 099-477. Montréal.

Falzon, P., and Mollo, V. (2009). Para uma ergonomia construtiva: as condições para um trabalho capacitante. *Laboreal*, 5(1).

Foray, D. (2009). *Economie de la connaissance.* Paris: La Découverte.

Garrigou, A., Daniellou, F., Carballeda, G., and Ruaud, S. (1995). Activity analysis in participatory design and analysis of participatory design activity. *International Journal of Industrial Ergonomics*, 15, 311–327.

Hubault, F., and Bourgeois, F. (2004). Disputes sur l'ergonomie de la tâche et de l'activité, ou la finalité de l'ergonomie en question. *Activités*, 1(1), 34–53. Retrieved from http://www.activites.org/v1n1/vol1num1.book.pdf.

Ijzelenberg, W., Molenaar, D., and Burdorf, A. (2004). Different risk factors for musculoskeletal complaints and musculoskeletal sickness absence. *Scandinavian Journal of Work, Environment and Health*, 30(1), 56–63.

Kausto, J., Miranda, H., Pehkonnen, I., Heliövaara, M., Viikari-Juntura, E., and Solovieva, S. (2010). The distribution and co-occurrence of physical and psychosocial risk factors for musculoskeletal disorders in a general working population. *International Archives of Occupational and Environmental Health*, 84(7), 773–788.

Kennedy, C. A., Amick, B. C., Dennerlein, J. T., Brewer, S., Catli, S., Williams, R., and Rempel, D. (2010). Systematic review of the role of occupational health and safety interventions in the prevention of upper extremity musculoskeletal symptoms, signs, disorders, injuries, claims and lost time. *Journal of Occupational Rehabilitation*, 20, 127–162.

Krause, N., Burgel, B., and Rempel, D. (2010). Effort-reward imbalance and one-year change in neck-shoulder and upper extremity pain among call center computer operators. *Scandinavian Journal of Work, Environment and Health*, 36, 42–53.

Kristensen, P. (2006). Prevention of disability at work. *Scandinavian Journal of Work, Environment and Health*, 32(2), 89–90.

Kuorinka, I. (1997). Tools and means of implementing participatory ergonomics. *International Journal of Industrial Ergonomics*, 19, 267–270.

Kuorinka, I. (1998). The influence on industrials trends on work-related musculoskeletal disorders. *International Journal of Industrial Ergonomics*, 21, 5–9.

Laing, A. C., Cole, D. C., Theberge, N., Wells, R. P., Kerr, M. S., and Frazer, M. B. (2007). Effectiveness of a participatory ergonomics intervention in improving communication and psychosocial exposures. *Ergonomics*, 50(7), 1092–1109.

Lièvre, P., and Coutarel, F. (2013). Sciences de gestion et ergonomie: pour un dialogue dans le cadre d'une économie de la connaissance. *Economies et sociétés*, K, 22, 1, 123–146.

Nathanael, D., and Marmaras, N. (2008). On the development of work practices: a constructivist model. *Theoretical Issues in Ergonomics Science*, 9(5), 359–382.

National Research Council. (2001). *Musculoskeletal disorders and the workplace: low back and upper extremity musculoskeletal disorders.* Washington, DC: National Academy Press.

Neumann, W. P., Eklund, J., Hansson, B., and Lindbeck, L. (2010). Effect assessment in work environment interventions: a methodological reflection. *Ergonomics*, 53(1), 130–137.

Ostergren, P. O., Hanson, B. S., Balogh, I., Ektor-Andersen, J., Isacsson, A., Orbaek, P., Winkel, J., and Isacsson, S. (2005). Incidence of shoulder and neck pain in a working population: effect modification between mechanical and psychosocial exposures at work? Results from a one year follow up of the Malmo shoulder and neck study cohort. *Journal of Epidemiology and Community Health*, 59, 721–728.

Petit, J., Dugué, B., and Daniellou, F. (2011). L'intervention ergonomique sur les risques psychosociaux dans les organisations: enjeux théoriques et méthodologiques. *Le Travail Humain*, 4, 391–410.

Punnett, L., and Wegman, D. H. (2004). Work-related musculoskeletal disorders: the epidemiologic evidence and the debate. *Journal of Electromyography and Kinesiology*, 14, 13–23.

Rabardel, P., and Béguin, P. (2005). Instrument mediated activity: from subject development to anthropocentric design. *Theoretical Issues in Ergonomics Science*, 6(5), 429–461.

Rivilis, I., Van Eedr, D., Cullen, K., Cole, D. C., Irvin, E., Tyson, J., and Mahood, Q. (2008). Effectiveness of participatory ergonomic interventions on health outcomes: a systematic review. *Applied Ergonomics*, 39, 342–358.

Roquelaure, Y. (2008). Workplace intervention and musculoskeletal disorders: the need to develop research on implementation strategy. *Occupational and Environmental Medicine*, 65, 4–5.

Sartre, J. P. (1943). *L'être et le néant*. Paris: Gallimard.

Schwartz, Y. (2010). Quel sujet pour quelle expérience? *Travail et Apprentissages*, 6, 11–24.

Simonet, P. (2011). L'hypo-socialisation du mouvement: prevention durable des troubles musculo-squelettiques chez des fossoyeurs municipaux. Unpublished doctoral dissertation, CNAM, Paris.

Sznelwar, L. I., Mascia, F. L., and Bouyer, G. (2006). L'empêchement au travail: une source majeure de TMS? *Activités*, 3(2), 27–44. Retrieved from http://www.activites.org/v3n2/LAERTE.pdf.

Tuncel, S., Genaidy, A., Shell, R., Salem, S., Karwowski, W., Darwish, M., Noel, F., and Singh, D. (2008). Research to practice: effectiveness of controlled workplace interventions to reduce musculoskeletal disorders in the manufacturing environment – critical appraisal and meta-analysis. *Human Factors and Ergonomics in Manufacturing*, 18(2), 93–124.

Van Rijn, R. M., Huisstede, B. A. M., Koes, B. W., and Burdorf, A. (2010). Associations between work-related factors and specific disorders of the shoulder – a systematic literature review. *Scandinavian Journal of Work, Environment and Health*, 36, 189–201.

Wells, R. (2009). Why have we not solved the MSD problem. *Work*, 34, 117–121.

Wilson, J. R., and Haines, H. M. (1997). Participatory ergonomics. In G. Salvendy (Ed.), *Handbook of human factors and ergonomics* (pp. 490–513). New York: Wiley.

Wisner, A. (1997). Aspects psychologiques de l'anthropotechnologie. *Le Travail Humain*, 60(3), 229–254.

chapter thirteen

Design projects as opportunities for the development of activities

Flore Barcellini, Laurent Van Belleghem
and François Daniellou

Contents

Activity ergonomics has developed, over the past 30 years or so, an approach to supporting design projects that combines ergonomic work analysis, a participatory approach and the simulation of work (Daniellou and Rabardel, 2005). It can be reported that this approach, when specific conditions are present, contributes to the development of activities, and not just to designing the solutions that are expected in the project. It thus allows the workers to take ownership of these solutions and to implement them, but also their control by other stakeholders in the company. This contributes to strengthening the sociotechnical system and the social relations as a whole.

This constructive dimension is not just a positive effect induced by this approach; it should be considered a driving force of the ability of men and women to cope with changes in their work situation, when they can actively contribute to designing this situation. The simulation of work plays a key part in this process.

The constructive dimension that is considered here focuses notably on the following:

- The development of activity and skills in workers throughout project management, allowing these workers to partially control these situations before they are even deployed.
- The development of designers' activity, through an anticipated confrontation between their proposals and the ongoing work in the real world, in the course of the design process itself.
- The development of the decision-making function, which is often made up of a set of stakeholders (company management, project management, human resources, etc.). These stakeholders must take on both a hierarchical role with respect to the populations involved in the project and the role of a contracting authority with respect to the designers.
- The development of the activity of staff representative institutions, who may find in the simulation of work a means to shift and restructure social relations.

This development does not operate in the same way for each of the agents taken individually. It is created in the meeting of 'worlds' (Béguin, 2003; Béguin, this volume), which this approach proposes, and which provides input for mutual learning between these stakeholders.

We propose here that this constructive dimension should become a clearly stated goal for ergonomic interventions in project management.

We first remind the reader of some perils of projects conducted with little reference to work in the real world. Following this, we describe the principles of an ergonomic approach to the conduct of design projects that has been developed in activity-centred ergonomics since the 1980s. Finally, we present:

- The known effects of simulations on the development of the activity of each of the stakeholders and on their relations with one another.
- An argument in favour of a position emphasizing an ergonomic approach of conducting projects as a constructive process.
- The need to pursue the evolution of the initial project of ergonomics, which focused on fitting the job to the human being, broadening this goal to the development of activities.

Perils of a poor integration of real-world work in project management

Numerous investment or reorganization projects lead to disappointing results, whose usual symptoms are delays in implementation and failure to complete the project within the budget because of the subsequent need for adjustments or the difficulties of workers to master the new system, an insufficient rate of operation (Wisner and Daniellou, 1984), long delays before achieving the target mode of operation in terms of quantity and quality, and sometimes, serious accidents.

An analysis of these malfunctions often reveals dual failures on the part of project management:

- On the one hand, the structure of the project itself is often at fault: a weakness in the political management of the project and in the definition of project goals; a weak presence of operations managers in the projects; supervision of the projects by engineering approaches that focus on the technical dimensions of work and underestimate the importance of aspects related to the population of workers, the organization of work and training; a lack of regular interactions between the definition of the will behind the project (contracting authority) and the search for solutions (main contractor) (Martin, 2000); the partial and late nature of information and the consultation of the staff representative institutions; and a very late discovery of the project by the people who will ultimately have to act in the new system.
- On the other hand, ergonomists have highlighted the ineffectiveness with which human work is taken into account in design decisions. The work that takes place in the organizations that predate the project is often approached only as a set of prescribed tasks. The regulations that male and female workers implement to cope with variability are left out, which may, for example, lead to designing overly simplistic automation schemes that are incapable of dealing with situations of variability (Daniellou, 1987). The future work, which will take place in the new system, is also approached through a set of prescribed procedures, with the assumption that work will merely involve the execution of these procedures. The constraints and leeway related to work activity, the consequences on health and the quality of production are poorly addressed.

In other words, everything happens as if the prescribers ignored, in their reflection about the upcoming evolutions, those evolutions that are related to the activity. Figure 13.1 summarizes the course of project management with and without consideration of real-world work, and the hindrance this may cause for the development of activities. The initial

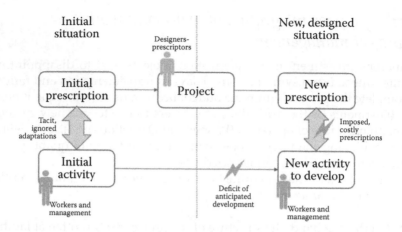

Figure 13.1 Failures in the development of activities in a mode of project management with no inclusion of real-world work.

situation is characterized by a kind of articulation between a prescription system and work activities, which mutually influence one another (Rabardel and Béguin, 2005; Guerin et al., 2006). By ignoring this connection, designers define a new system of prescriptions that, once it is implemented, is supposed to be mechanically 'executed'. The perils we have noted above quickly emerge. Two failures of design projects may then explain the emergence of tensions related to these perils:

- The inadequacy of the prescribed system: The new system of prescriptions (tools, spaces, rules of organization, etc.) does not take enough into account the rules structuring activity and its variability. The result is regulations that are costly to workers.
- A lack of development: The activities that are useful to operating the new system are not well developed enough at the time of system implementation. It is often expected that this development will occur when experimenting with the new system, but the possibilities of such a development are often hindered by the mismatch of the new system that has been designed with the logic underlying the activity. Tensions become long lasting.

In the face of the inadequacy of prescribed systems, activity ergonomics (Garrigou et al., 1995; Theureau, 2003; Daniellou and Rabardel, 2005) has developed since the 1980s approaches to design, followed by approaches to project management, that aimed to foster better interactions between project stakeholders, a better integration of existing work, and an anticipation of future work. The section below will present an actualized summary of this work.

Project management in ergonomics: An approach aiming to enrich the project by taking into account real-world work

In activity ergonomics, design projects are concerned less with defining the characteristics of artifacts (products, tools, spaces, workstations, etc.) and more with defining the features of the work situations in which these artifacts will be present. Indeed, in a work situation, activity is deployed within a space that is framed by a set of prescriptions: tasks to be carried out, work spaces, hardware and software, organizational structure (numbers of workers, types of contracts, formal allocation of tasks, schedules, rules, etc.), the training programs that are made available, etc. The design of the work situation then focuses on defining these various components and the relations between them, in order to allow the deployment of an activity that is effective and protective of the health of workers.

Project management aims to frame this design project within the company*, and relies on an approach that is

- Defined at the start of the project.
- Guided by the will to design or transform one or several work situation(s).
- Socially situated, since it involves a collective of stakeholders, each bearing different perspectives.
- Structured by an organization and a constrained temporal and financial framework.

This approach suggests carrying out a reversal in the classical relationship between ergonomists and designers. Ergonomists should not focus just on providing input, in the form of recommendations, to processes for designing artifacts that are mastered by the designers alone. They should instead contribute to the implementation of a global structuring approach (see Theureau, 2003, for a synthesis) within the company, in order to move from a project that is managed based on technical criteria to one that is managed based on existing and future work. This approach makes it possible not only to contribute to the design of a high-quality work system, but also to enrich the very goals of the project. Design decisions are then informed by trade-offs between the various dimensions of performance (human, technical, economic, etc.), the connection between them and the stakes related to health.

This approach must rely on an analysis of the project and of work activities (carried out by ergonomists), on the implementation of a structured,

* The term *company* refers here to any employing institution (industry, hospital, administration, etc.).

participatory, joint approach (fostered by ergonomists), and on the conduct of simulations of work – supported by ergonomists – that make it possible to project oneself in the probable future activity, the formalization of the results of the simulations, directed toward the project stakeholders (carried out by ergonomists, collaborating with the project stakeholders), and support for the project (performed by ergonomists) until it starts.

The stakeholders that must absolutely be involved in this approach of participatory design are as follows:

- *Decision-makers,* including a range of stakeholders (company managers, project managers, human resources). These are the bearers of the intention behind the project and often exert a twofold function. They are part of a functional relationship with prescribers (for example, as a contracting authority) as well as a hierarchical relationship with their employees, whose situation they elect to change. Because of this, they play a crucial role in defining the trade-offs between the expected goals of the project and the effects of that project on the real-world work.
- *Workers,* whose activity will be transformed in the situations covered by the project (including middle management).
- *Designers,* but also, more broadly, any function, whether internal or external to the company, that is involved in prescribing work (engineering departments and companies, organizational consultancy firms, etc.), which we will name prescribers.
- *Staff representative institutions,* which must find a genuine place in the system being implemented.

This approach is modelled in Figure 13.2. It comprises three main stages (analyzing, simulating and supporting) that we will describe below. The simulation of future activity is the heart of this approach.

Analyzing: Constructing knowledge about the project and work in the real world

Project analysis focuses on the prime intentions behind the project, and on its stakes (economic stakes, production stakes, stakes related to work activity), whether these are made explicit or not, and on the project structure that has been implemented. This project structure ties together a will related to the future, borne by the contracting authority, and the search for solutions, borne by the main contractor, on identifying the population involved in future work situations. Project analysis also covers the data related to health and system performance. This analysis makes it possible to formulate a project diagnosis that is geared toward the decision-makers, and to contribute to structuring and redefining project goals.

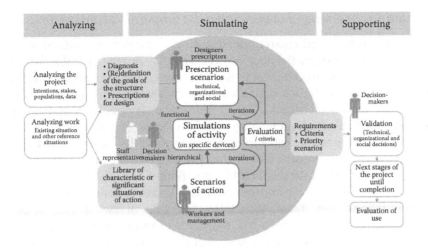

Figure 13.2 (Updated) approach to project management, as proposed by activity ergonomics.

Ergonomic work analysis is the first stage in the approach of project management. Its goal is to produce knowledge that is related to work, that will be useful to inform project choices (e.g. helping enrich the goals, the structure and the early design choices) and to the continuation of the process (handing down key elements to designers and conducting simulations). Work analysis is conducted in any work situation that is said to be a *reference situation*, whose determinants – whether technical, organizational or social – are relevant with respect to the initial situation, or to the future work situation. This analysis may produce many outputs: it contributes to enriching the project, but also aims to generate knowledge about work that is necessary for continuing the process. The formalization of this knowledge is directed:

- Toward the decision-makers, who are in a position to make project goals evolve based on the insights of the analysis.
- Toward designers and prescribers, by formalizing references to help construct the early design solutions. We will give to these references a status of *prescriptive scenarios*.
- Toward the *ergonomist*, through the construction of libraries containing characteristic situations of action (Garrigou et al., 1995; Theureau, 2003). These situations account for the variability of situations encountered by workers, making it possible to identify situations they may have to deal with in the future. They make it possible to construct 'action scenarios' that will be 'played out' during the simulations.

Simulating: Assessing and enriching the proposals of designers

Simulations aim, based on an understanding of the current state of work in the real world, to have the persons involved play out the probable future work (based on action scenarios) in the conditions that are imposed by the new prescriptive scenarios suggested by prescribers. Simulating work is a projective method (Daniellou, 2007) that makes it possible to anticipate the conditions in which an activity will be carried out, in a given set of conditions. It makes it possible to assess the proposals made by prescribers, guiding choices toward this or that prescriptive scenario – scenarios being represented by intermediary objects (Boujut and Blanco, 2003) – and fostering their progressive improvement in a series of iterative stages.

Simulations may take place following one of two main modes (Daniellou, 2007). They can be full scale, for example, by using a prototype: workers may then personally experience the improvements that may or may not be brought on by the new solution. This makes it possible, in turn, to bypass the defences that may lead stakeholders to think that it is impossible to improve upon the situation. They can also be a small-scale simulation (for example, a mock-up). In this case, the activity may be described verbally (one then uses the term *language-based simulation*). However, this kind of simulation runs the risk of simulating the sequence of tasks rather than the activity. The use of avatars to mediate the activity (Barcellini and Van Belleghem, 2014) may help with the 'game' of simulation, by allowing workers to embody the simulated actions, and contribute to realistic descriptions of the activity. The simulation provides some input for the dialogue taking place between workers and prescribers, and allows the construction of negotiated design solutions (e.g. Détienne, 2006; Béguin, 2003; Béguin, this volume) and trade-offs that experience shows may be innovative ones. This dialogue takes place at the discretion of decision-makers (connected with the staff representative institutions) who must open up or close down specific orientations to search for solutions, depending on the goals pursued by and the means available for the project.

Supporting: 'Converting the try' of simulations

Simulations are at the heart of the ergonomic approach of design. However, implementing a simulation is not enough in itself to act on the work situation that is being constructed. The chosen prescription scenarios must also be validated by the project's decision-making authority. They must also be genuinely implemented and deployed upon project launch. This has several consequences in methodological terms. Approval by the decision-making authority implies that the choices that are made, the trade-offs that are performed in design, should be justified with respect to the stakes of the project. Decision criteria need to be 'sketched out' in the simulation sessions. The concrete implementation of the prescription scenarios implies translating the results of simulations in the form of requirements that can be taken over by the designers. These formalized requirements are an essential resource for the work of designers. They make it possible to move forward in the design of the future system, in its concrete realization, until the beginning of the project. The approach supports this development by implementing iterative, increasingly detailed simulations that make it possible to refine the definition of the system to be designed, until it is launched.

The approach we have presented here has been implemented many times since it was first formalized. Today, it focuses not just on the design of technical systems, but also on the design of work organizations (Petit, 2008; Petit et al., 2011; Petit and Dugué, 2012; Barcellini and Van Belleghem, 2014; Coutarel and Petit, this volume). On these various occasions, we were able to notice forms of activity development, notably in the simulation stages. From our point of view, these forms of development compensate for the lack of development that we highlighted in the introduction of this chapter. Whereas until now, this development has been viewed as a side effect of this approach, we suggest this development might be seen as a goal of this approach in itself.

Design projects as opportunities for the development of activities

Implementing the approach outlined above produces effects that frequently go beyond the initial goals of the project. In particular, one can observe that equipping the participation of stakeholders to this approach will contribute both to the design of the future situation and to the development of their own activity, in the course of the approach. It also contributes to the development of the activities of the other stakeholders involved in this approach: decision-makers, designers and staff representatives. Because of this, when the conditions are adequately met, this approach strengthens these functions and the professional relations *between them*, both over the course of the project (cooperation in the project) and following its operational implementation (cooperation in work). One can point out that the success of the project relates to the relevance of the choices made during the project, as much as to the quality of the developmental processes it helped support.

We will describe below in detail the development processes that we have observed, depending on the type of stakeholder.

Development in the activity of workers

The approach we propose locates a particular activity situation, the *situation of simulation* (Béguin and Pastré, 2002), between the former situation of activity and the new situation that is being designed. This transition situation provides an opportunity for the development of activities of future users of the system (in particular, various workers and middle managers) before the project is implemented, whilst contributing to its design. Figure 13.3 summarizes the progression of this situation of simulation.

As we have pointed out, simulations aim to have workers play out their probable future work based on the prescription scenarios suggested by designers and on the action scenarios derived from the analysis of existing situations. It is both the prescription scenarios and the activity that are tested. In the first stage, this trial always reveals difficulties, or even dead ends and disagreements (Béguin, this volume), in the realization of the simulated activity. These difficulties do not have any severe consequences, since they emerge before the solutions are produced (this justifies the use of the term *scenario*, since scenarios allow some evolutions). On the contrary, the early revelation of these difficulties – contrary to what happens all too often in classical design processes – reveals some shortcomings that may lead, later on, to difficulties in system operation that could be avoided by altering the prescription scenarios. However, these difficulties may also reveal a need for the development of activities, so that workers might better take ownership of the principles of prescription that might prove to be relevant. Simulations must here, once again, allow testing of the forms of appropriation. The scenario might be adapted to foster this process of development. When carried out in this way, this

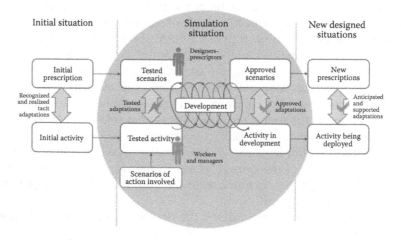

Figure 13.3 The situation of simulation, a mould for the development of activities.

approach leads to designing a prescription system based on the level of development that it allows.

This dual motion of development of prescription and development of activity echoes the process of instrumentalization/instrumentation proposed by Rabardel (Rabardel and Bourmaud, 2003; Rabardel and Béguin, 2005; Bourmaud, this volume) that, in turn, draws inspiration from the process of accommodation/assimilation proposed by Piaget.

Iterations in this dual movement, which rest on a wide range of characteristic situations of action, lead to the elaboration of a prescription scenario that has been sufficiently stabilized through successive trials. But they also lead to the elaboration of new forms of activity, which are tried out by workers taking part in the approach even before the process is launched in concrete terms. When this occurs, the activity that has started developing during the simulations may be deployed in the new system, continuing its development. It is quite common, just before implementing a new installation, software program or organization, to hear the workers who took part in the approach say they 'cannot wait to be there' and express their confidence in the success of the project. This finding is in sharp contrast with numerous forms of project management where forms of resistance and mistrust are often observed. However, the simulation focuses on several prescription scenarios – not just the scenario that will ultimately be selected. Because of this, it covers a broad space of exploration and experimentation, describing ways of doing work, and opening them up for debate or even controversy – not just between workers and prescribers, but also between the workers themselves. Indeed, it is this very kind of debate, which will focus notably on the criteria involved in the quality of work, that Clot (2011) calls for in order to set the foundations for 'well-doing' – something without which no 'well-being' would be possible.

Thus, simulations contribute to the construction of a rich experience of the various scenarios explored, gradually expanding the range of possible 'professional gestures' (Clot, 2011, p. 102). Even if the activity developed in simulations will continue to develop in real-world situations, it may sometimes cover a broader range of possibilities than what will be required in the new situation. Indeed, the situation of simulation comprises numerous trials and errors that are subject to collective arbitrage, and that continue to provide input for the activity in development.

To avoid discrepancies between the workers who took part in simulations and the other workers, it is necessary to design a complementary training system that, once again, will rely on the implementation of simulations of work, but focusing only on the scenario that has been tested out and approved. From this training system, one can expect the development of new activities focusing on the use of the future system, even before its operational implementation.

Development in the activity of designers

Designers are invited to take an active part in the simulation. They are expected to suggest some initial prescription scenarios (and not just a single solution) based on the reference materials produced by the ergonomic diagnosis, to attend the simulations carried out by the workers, to discuss and argue their evaluation criteria, and to help make scenarios evolve toward a better integration of the activity in development. On this occasion, one can observe some development in the activity of designers.

Designers, who are most often involved in a process with a strongly technical view that is focused only on the material or procedural aspects of the system to be designed, are confronted with the activity of the workers and with its possibilities for development. When made visible through simulation, this activity may, at first, appear as an additional constraint to them, which would be difficult to ignore. But it quickly becomes the central issue of design when it emerges as the very condition of system operation.

At this point, a reversal will occur. Whereas they used to be constraints, the increased requirements related to the recognition of work activity in the real world become a resource for designers. By providing designers with tools to imagine new solutions to respond better to the debate between criteria that begins between participants (for example, productivity requirements and quality requirements), they constitute a source of innovation for them (in the sense that an innovation is an appropriation). If the task, as Clot (1999) points out, is a 'cooled-down product' of the activity of designers, there is a key stake in testing this activity, as long as it is still 'hot' with respect to the activity of the workers that is still in development. It is in this mould, formed by participatory design and made 'white hot' by the simulation of future work, that the mutual learning (Béguin, 2003; Béguin, this volume) between designers and workers takes all of its strength, and contributes to the development in the activity of designers.

Initiating this process implies that the activity of design should be distributed and that the designers agree to share part of the design, notably regarding:

- The elaboration of prescription scenarios, for which each participant (workers, managers, decision-makers, staff representatives) can be a source of propositions.
- The elaboration of criteria for the evaluation of the design system. These will no longer refer only to the technical validity of the system, but will be completed by criteria that are related to the development of an efficient and meaningful work activity.

It is quite surprising, from this point of view, to note that the development of the activity of designers seems to be dependent upon being 'dispossessed' from one part of that activity. Here, the simulation situation seems to offer an opportunity for designers to reflect upon their own practice (Schön, 1983), helping them guide their activity toward a greater relevance of the designed system, with the possibility of moving them away from their initial models. This may contribute to the renewal of the profession of engineering, which has fallen prey to the misgivings of project management (Charue-Duboc and Midler, 2002).

Development in the activity of decision-makers

The approach contributes to the development of the decision-making function. It allows decision-makers to realize the primacy – and therefore the liability – of the function of contracting authority, which defines the will related to the future, compared with contractors, whose function is to identify solutions to implement this will. Discrepancies between what is wished for and what can be done require trade-offs, through adjustments of goals and resources. This preeminence must apply to all aspects – technical, organizational, training related – over the entire duration of the project, which implies organizing the decision-making activity with the following:

- On the one hand, a team of managers representing various logics that are all vital to the development of the company (finances, marketing, human resources, quality, safety, environment, etc.)
- On the other hand, a permanent representation of this collective by a project manager who deals with the everyday interactions with contractors

The development of decision-making activity implies, notably, strengthening the acknowledgement of the diversity of logics that need to be taken into account, of the contradictions between these logics, and of the interest in the collective construction of trade-offs within the management team. By confronting technical and organizational prescriptions with simulations of real-world work, the ergonomic approach contributes to implementing a restoring force grounded in the real world, and to avoiding the construction of 'managerial defences' such as 'it'll work regardless'. It also helps reveal possible hidden expenses and conflicts between criteria. But it may also contribute, sometimes, to opening up new perspectives for organization and management, which would have appeared to be incongruous or inaccessible, had simulations not shown them to be viable solutions.

This point may lead decision-makers to alter their strategy for announcing future projects: rather than waiting for the project to be stabilized before making an announcement, they may consider the early announcement of projects that are still uncertain as a positive opportunity to enrich those projects by debating them in the light of the possibilities available to the activity.

An ergonomic approach to project management also contributes to reinforce the workers' and middle managers' 'right to play' as participants in the design process. The discovery by decision-makers that their contribution has led to avoiding major mistakes may facilitate the implementation of new management practices. This is also the case with the relations with staff representative institutions.

Development in the activity of staff representative institutions

The ergonomic approach described above also constitutes a trial and an opportunity for staff representative institutions. It is a trial because it may come in to question a culture of representation by delegation, where staff representatives may think of themselves as the only legitimate bearers of the point of view of employees on the conditions of work. It is also an opportunity because the results of activity analysis cast a light on employees not just as being constrained by their work environment, but also as being fully committed to the pursuit of what they consider to be quality work (Chassaing et al., 2011; Petit and Dugué, 2012), and as bearers of unsuspected knowledge. In some cases, this reversal has led trade unions to explicitly alter their practices regarding relations with employees, seeking to ground these practices in an understanding of work.

The approach may contribute to the development of the activity of staff representative organizations when there are major organizational stakes at play, by providing two main tools: a detailed understanding of the existing activity and a reflection on the future in terms of consequences on future work. A structured process of informing choices can thus develop in lieu of previous ritual confrontations (Dugué, 2008; Guerin et al., 2006). 'Tracking' design decisions over the course of them being formalized also allows staff representative institutions, if need be, to remind other stakeholders of their importance in case they had been left out of the later stages of project realization.

This development of the activity of staff representative organizations around the issue of work does not imply that they should not take into account more dimensions than are taken into account in ergonomic simulations: employment, pay, status, etc. In some cases, it can be noted that this 'macro' aspect is nourished by the reflection about work that has been developed in the project. Such is the case, for example, when negotiations regarding the size of the workforce rely on the action scenarios used in the simulations to ensure that the staff will be in sufficient numbers to cope not only with normal situations, but also with incidental ones.

Conclusion: From the adaptation of work to human beings to the development of activity

Acknowledging the development of activities as a part of the ergonomic approach to project management (and more generally, as part of any work situation) is one aspect of an evolution of the foundational project of ergonomics, whose goal was 'adapting work to human beings'. This view was initially reflected in the original orientations of the Ergonomics Research Society in 1949, under the heading of 'fitting the job to the worker', through the search for a match between work and the psychological and physiological characteristics of the human being. This orientation is also reflected in what Hubault and Bourgeois (2004) have called 'the ergonomics of tasks', with the development of ergonomic guidelines that could be applied directly to the design of work systems*.

Around the same time, 'activity ergonomics' was developed (mainly in French-speaking countries, but also in Scandinavian or South American ones), which highlighted the active contribution of workers to the completion of tasks, considering the inescapable variability of real-world situations. This activity ergonomics, which was originally focused on the understanding of work, changed its orientation in the 1980s toward taking into account work activity in project management, as described above. The stake of this approach is to spaces for future activity (Daniellou, 2007) that leave some leeway for workers to manage their activity – or even possibilities for the continuation of design in use (Rabardel and Béguin, 2005).

The prospect of 'constructive ergonomics', then, is to broaden the concept of adaptation or fit to that of the development of activity. The goal of ergonomists is to contribute to the design of work situations that will serve as a starting point for the development of the activity of the men and women who are the stakeholders of the project. To allow this development to start early, in the design of work situations, and in so doing, turning the criterion of the construction of experience (through the simulation of work), a criterion to assess the solutions produced by the prescriber-designers may prove to be a strategic choice for ergonomists.

This prospect would, no doubt, call for strengthening research programs, concerning both the learning that is involved in the various stakeholders in project management (including the ergonomist) and the practical methods and practices for intervention that may support the development of their activities.

* This dynamic approach was developed in the United States under the term *human factors*, and in the United Kingdom under the term *ergonomics*. The convergence between the two terms was reflected in the change in the name of the Human Factors Society, in 1992, to the Human Factors and Ergonomics Society, and by the change in the name, in 2009, of the British Ergonomics Society to the Institute of Ergonomics and Human Factors.

References

Barcellini, F., and Van Belleghem, L. (To appear). Organizational simulation: issues for ergonomics and for teaching of ergonomics' action. In O. Broberg et al. (Eds.), *Proceedings of Human Factors in Organizational Design and Management*. XI Nordic Ergonomics Society Annual Conference, 46. Santa Monica, CA: IEA Press.

Béguin, P. (2003). Design as a mutual learning process between users and designers. *Interacting with Computers*, 15(5), 709–730.

Béguin, P., and Pastré, P. (2002). Working, learning and designing through simulation. In S. Bagnara, S. Pozzi, A. Rizzo, and P. Wright, (Eds.), *Proceedings of the 11th European Conference on Cognitive Ergonomics: Cognition, Culture and Design* (pp. 5–13). Rome: Instituto di scienze e technologie della cognizione consiglio nazionale delle ricerche.

Boujut, J. F., and Blanco, E. (2003). Intermediary objects as a means to foster co-operation in engineering design. *Journal Computer Supported Cooperative Work*, 12, 205–219.

Charue-Duboc, F., and Midler, C. (2002). Concurrent project management and engineering departments. *Sociologie du Travail*, 44(3), 401–417.

Chassaing, K., Daniellou, F., Davezies, P., and Duraffourg, J. (2011). *Prévenir les risques psychosociaux dans l'industrie automobile: elaboration d'une méthode d'action syndicale*. Recherche action. Emergences-CGT-Ires.

Clot, Y. (1999). *La fonction psychologique du travail*. Paris: PUF.

Clot, Y. (2011). *Le travail à cœur: pour en finir avec les risques psycho-sociaux*. Paris: La découverte.

Daniellou, F. (1987). Automatize what: facts or fiction? In A. Wisner (Ed.), *New techniques and ergonomics*. Paris: Hermès Publications.

Daniellou, F. (2007). Simulating future work activity is not only a way of improving workstation design. *@ctivités*, 4(2), 84–90. Retrieved from http://www.activites.org/v4n2/v4n2.pdf.

Daniellou, F., and Rabardel, P. (2005). Activity-oriented approaches to ergonomics: some traditions and communities. *Theoretical Issues in Ergonomics Science*, 6(5), 353–357.

Détienne, F. (2006). Collaborative design: managing task interdependencies and multiple perspectives. *Interacting with Computers*, 18(1), 1–20.

Dugué, B. (2008). Les paradoxes de la participation du CHSCT dans la conduite des projets de conception. In P. Negroni and Y. Haradji (Eds.), *Ergonomie et conception, 43e congrès de la SELF* (pp. 49–53). Toulouse: Octarès.

Garrigou, A., Daniellou, F., Carballeda, G., and Ruaud, S. (1995). Activity analysis in participatory design and analysis of participatory design activity. *International Journal of Industrial Ergonomics*, 15, 311–327.

Guerin, F., Laville, A., Durrafourg, J., Daniellou, F., and Kerguelen, A. (2006). *Understanding and transforming work: the practice of ergonomics*. Lyon, France: ANACT Network Editions.

Hubault, F., and Bourgeois, F. (2004). Disputes sur l'ergonomie de la tâche et de l'activité, ou la finalité de l'ergonomie en question. *Activités*, 1(1), 34–53. Retrieved from http://www.activites.org/v1n1/vol1num1.book.pdf.

Martin, C. (2000). *Maîtrise d'ouvrage, maîtrise d'œuvre, construire un vrai dialogue. La contribution de l'ergonome à la conduite de projet architectural*. Toulouse: Octarès.

Petit, J. (2008). Ergonomic intervention as learning dynamics in organizational design. In L. I. Sznelwar, F. Mascia, and U. B. Montedo (Eds.), *Proceedings of human factors in organizational design and management* (pp. 591–596). São Paulo: Editora Blücher.

Petit, J., and Dugué, B. (2012). Psychosocial risks: acting upon the organisation by ergonomic intervention. *Work*, 41(Suppl. 1), 4843–4847.

Petit, J., Dugué, B., and Daniellou, F. (2011). L' intervention ergonomique sur les risques psychosociaux dans les organisations: enjeux théoriques et méthodologiques. *Le Travail Humain*, 74(4), 391–409.

Rabardel, P., and Béguin, P. (2005). Instrument mediated activity: from subject development to anthropocentric design. *Theoretical Issues in Ergonomics Science*, 6(5), 429–461.

Rabardel, P., and Bourmaud, G. (2003). From computer to instrument system: a developmental perspective. *Interacting with Computers*, 5(5), 665–691.

Schön, D. A. (1983). *The reflective practitioner: how practitioners think in action.* London: Temple Smith.

Theureau, J. (2003). Course of action analysis and course of action centered design. In E. Hollnagel (Ed.), *Handbook of cognitive task design* (pp. 55–82). London: CRC Press.

Wisner, A., and Daniellou, F. (1984). Operation rate of robotized systems: the contribution of ergonomic work analysis. In H. W. Hendrick and O. Brown Jr (Eds.), *Human factors in organizational design and management* (pp. 461–465). Amsterdam: Elsevier Science Publishers B.V.

chapter fourteen

Reflective practices and the development of individuals, collectives and organizations

Vanina Mollo and Adelaide Nascimento

Contents

Current organizations, in becoming dynamic and immaterial, must be capable of responding to the needs of the market through their ability for adaptation, improvement and innovation (Devulder and Trey, 2003), and not just through their productive ability. Although in the past, the stability of work systems led to learning by repetition, and the development of skills and knowledge, in both individuals and collectives it was a more or less random outcome of work; today, being able to learn from singular cases and develop individual and collective knowledge is a requirement of performance.

We assume that each work activity has a *productive dimension*, directed toward the worker and the objects of his or her task, which involves transforming the real world (in its material, symbolic and social aspects), and a *constructive dimension*, as the worker transforms himself by transforming the real world (Samurçay and Rabardel, 2004; Delgoulet and Vidal-Gomel, this volume). The time of productive activity follows the course of the activity itself, whereas the time of constructive activity takes place in another span, a longer duration of time that goes far beyond action and focuses on the development of the individual. In this sense, skills display a dynamic nature, as the results of action are reflected within activity through 'evolutions in representations of situations and in their management' (Weill-Fassina and Pastré, 2004, p. 221, our translation).

These evolutions of representations are derived, amongst other things, from situations of reflectivity that make it possible to distance oneself from the action. This space-time, beyond that of productive activity, allows workers to prepare later activities, to exchange resources with colleagues, and to take a step back from what they have just done. Based on these principles, individuals learn, both individually and collectively, using the knowledge derived from the results of their own activity. The rules and knowledge constructed in this way may become an effective tool to elaborate action, and also benefit organizations.

The development of skills thus combines learning in action with learning in the analysis of action: 'it is the connection between these two moments that is probably characteristic of the construction of professional experience' (Pastré, 2005, p. 9, our translation).

The goal for ergonomics is no longer, as it was in the past, to highlight the knowledge and know-how that is developed through practice. It must support this motion using appropriate reflective methods. As Amable and Askenazy (2005) have pointed out, it is at least as important to learn how to learn as it is to learn (see Six-Touchard and Falzon, this volume).

Such methods pursue a twofold goal of understanding and action. They aim to provide each person with a better visibility and understanding of the work of others, to homogenize practice or to construct the envelope of acceptable practice, thus fostering the emergence of a collective culture (for safety, quality, etc.). The issue is to develop individuals and collectives at the same time.

In other words, the goal is to develop the enabling potential of organizations (Falzon, 2005, 2007) 'so that they might contribute simultaneously and sustainably to the well-being of employees, the development of skills, and the improvement of performance. Every organization has access to a more or less sizeable enabling potential. However, this potential is often underused, unknown, or little known, and sometimes even hindered by the organization' (Falzon and Mollo, 2009, our translation).

After having briefly defined reflective practice, we will present and illustrate several methods that allow supporting this practice. This will open the way for a few of the essential conditions (the 'golden rules') to ensure that reflective practice is indeed a constructive practice.

Reflective practice as a constructive practice

Developing the potential for action of individuals

The idea that action is a source for the development of knowledge is not a new one in ergonomics. Every work activity comprises a functional, or productive, dimension that is directed straight to the completion of a task, and a reflective or constructive dimension, whose object is the productive activity and which transforms it back, by transforming the workers (Falzon, 1994; Samurçay and Rabardel, 2004). These two dimensions cannot be separated and feed off one another.

Reflective activity implies a critical analysis of the activity 'either to compare it to a prescriptive model, to what one should or could have done differently, to what another practitioner might have done, or to explain and critique it' (Perrenoud, 2001, p. 31, our translation). It can be carried out at the same time as the activity (through the control and evaluation of action) through a 'reflective conversation with the situation' (Schön, 1983) or posterior to the activity. In either case, its effects go far beyond the temporal logic of the action in progress. Reflective activity makes it possible to construct knowledge and know-how that is destined to an eventual future use and thus contributes to facilitating the execution of a task or to improving performance (Falzon et al., 1996; Falzon and Teiger, 1995). Indeed, reflective practice makes it possible to draw lessons from past experience, through the analysis of what has been done, but also of what was not done, and what one was prevented from doing (Clot, 2008), what was done by others. Hence, it develops the potential of workers for action, their ability for trade-offs, making them increasingly able to respond to the variability of situations, either as individuals or as collectives.

As noted by Perrenoud (2001), one should make a distinction between reflective practice and episodic reflection: 'to move towards a true reflective practice, this must become an almost permanent position, which takes place within an analytical relationship with action, that would be relatively detached from the obstacles encountered or from disappointments.... Reflection is not limited solely to evocation, but relies on a critique, an analysis, on drawing relations with rules or other actions, whether imagined or conducted in an analogous situation' (p. 14, our translation).

We will use the term *reflective activity* to refer to this form of reflection on action, carried out outside of the immediate functional framework, allowing an individual or collective critical analysis of a single work situation or a family of situations.

In the next section, we describe the main characteristics of collective reflective activity.

Developing the potential for action of collectives

Collective reflective practice is closely related to 'debative' (Schmidt, 1991) or 'confrontative' (Hoc, 1996) cooperation. It rests on the confrontation of a group of professionals to the work activity of one or several of its members, whether or not they belong to the same field of expertise or occupy the same function in the hierarchy.

The goal of collective reflective practice is to learn from experience: 'beyond established knowledge, knowledge in action, know-how, malfunctions, "know-not-how" reveal knowledge that should be constructed and transferred. The goal is to know practices and to switch from knowledge in action to knowledge of action' (Gaillard, 2009, p. 154, our translation).

Collective reflective practice takes its place in the lineage of socioconstructivist theories, which emphasize the importance of social interactions in individual and collective learning, being notably close to the theory of sociocognitive conflict developed by Doise and Mugny (1981). According to this theory, social interactions are constructive in the sense that they introduce a confrontation between diverging views (Garnier, 2005). This confrontation triggers a realization in the individuals, as they discover points of view that differ from their own. As a result, this leads to questioning the points of view of each person, which may be a source of 'cognitive progress' (George, 1983) through the collective resolution of conflicts.

Hence, just like group deliberation (Manin, 1985; Urfalino, 2000), collective reflective practice is more than the adding up of individual reflections and has a dual effect, both individual and collective:

- An individual effect, because the perception of different points of view leads each agent to analyze them with respect to his or her own point of view, thus broadening, completing or altering it
- A collective effect, because confrontation allows the development of new knowledge and know-how

Collective reflective practice, beyond the fact that it offers – just like reflective practice at the individual level – possibilities for the development of skills, has two main interesting features.

On the one hand, it offers a space to support exchanges about work, which contributes to improve the effectiveness of the productive activity, either through the definition of a solution taking into account as many criteria as possible ('objectifying solutions or making them reliable' (Barthe and Quéinnec, 1999, our translation)) or by increasing the number of possible alternative solutions that are generated in the exchange (Clark and Smyth, 1993, cited by Hoc, 1996).

On the other hand, it makes it possible to collectively debate the variability of situations and how that variability may be processed, thus developing the ability of individuals and collectives to make trade-offs in order to respond to future situations, both new and unforeseen. In so doing, collective reflective activity may be a means to foster the production and the conservation of a work collective that, by reconstructing the rules to reduce conflicts between goals, forms a resource for the health of workers (Caroly, 2010; Caroly and Barcellini, this volume) and for overall system performance (Daniellou et al., 2011). We might also think that collective reflective activity develops mutual trust between workers via the development of a work collective.

In spite of these benefits, reflective practice often remains hidden, unacknowledged or even fought against because it is not 'immediately productive'. The goal for ergonomists, then, is to make it visible when it does exist, or to support its inception when it does not. In other words, the goal is for this constructive activity to become a productive activity within the organization. Several methods exist to achieve this, described in the sections below.

Methods for supporting reflective activity

A number of methods in ergonomics make it possible to encourage the development of a reflective activity. These methods aim to support a reflection grounded in action (the activity is the object of analysis), on action (individual or collective auto-analysis) and for action (to improve it, to construct new knowledge, and to act on the practice). The main feature of these methods is that they involve a realization by the worker, which is characterized by at least two factors (Mollo, 2004; Mollo and Falzon, 2004):

- Workers are located some distance away from the task environment. This allows them to concentrate on their knowledge and skills that they apply in the course of their activity.
- By becoming analysts of their own activity or that of other people, workers elicit what they actually do, why and how they do it. The goal is not just, therefore, to say what they know, but also to discover some implicit knowledge and other ways of doing the work.

Workers are therefore viewed both as workers and as analysts. This constitutes a starting point for reflective practice. This is obviously not new in the history of ergonomics (Teiger, 1993; Teiger and Laville, 1991). However, the two features noted above are at the heart of the methods to assist collective reflective practice.

The activity that is the object of reflection may be represented in various ways (films of the activity, narrations of work situations, reports on the observation of an activity, etc.). It is useful to draw a distinction, in terms of benefits and practical organization, between individual and collective methods of confrontation. To do this, we will rely on Mollo and Falzon (2004).

Individual auto-confrontation

Individual auto-confrontation consists in confronting an operator with his own activity. It makes it possible to access the logics underlying the activity and leads the workers:

- To realize their own know-how through the description of their own activity
- To elicit the logics that underlie this know-how, which are not necessarily conscious, but which become so in the process of elicitation

Individual auto-confrontation sometimes forms a crucial preliminary step for other forms of confrontation to latch on to.

Allo-confrontation

Allo-confrontation consists of confronting a worker with an activity that he or she carries out on a daily basis, but which is being performed by a colleague.

The expected benefits of using this method are as follows:

- A change of representation, resulting from the fact that the operators are purposely placed at a distance from their own point of view
- A realization of other forms in which the activity can be performed, leading workers to realize the features of their own activity in comparison to the activity of others
- A critical analysis of his or her own knowledge and know-how, compared with that of other workers
- The construction of new knowledge

Allo-confrontation may be individual, or it may be crossed. The first kind consists of confronting a worker with the activity of another worker, without that worker being present (but with that worker's approval). In the second kind of allo-confrontation, two workers comment on the activity of their respective colleague (this method is also known as 'crossed auto-confrontation'; Clot et al., 2000). From the point of view of the commenting worker, the features are the same as those described above. However, the worker whose activity is being commented on is confronted with the representation the other worker has of his activity, leading the former to better justify his knowledge and to elicit some aspects of his activity that might not have been developed had his colleague not intervened. Therefore, commenting on the activity of a colleague in the presence of that colleague profoundly changes the situation and its benefits.

Crossed allo-confrontation is one particular form of allo-confrontation that typically ends with an exchange between the two protagonists. Because of this, it brings additional benefits to the method – that is, the construction of new, shared procedures.

Collective confrontation

Collective confrontation is a form of collective reflective activity during which a group of workers comments on the activity of one or several of these workers. The collective is made up of workers who may or may not belong to the same field of expertise or to the same hierarchical function.

This method makes it possible to

- Elicit the representations of group members.
- Construct shared representations and shared knowledge. This construction is the result of sharing individual experiences, allowing mutual learning.
- Collectively evaluate the various modes of activity realization and the solutions derived from the confrontation.

The dynamics of the exchange lead the workers to elicit and assess their own knowledge and know-how with respect to those of the others, and to construct new knowledge and know-how. However, this process is not always visible, nor is it elicited: knowledge may be constructed without the worker sharing it with the collective.

Cost–benefit analysis of the methods presented

As we have pointed out above, confrontation methods may be used from the perspective of understanding or from a perspective of action (Mollo and Falzon, 2004). The first perspective is directed mainly toward the analyst. Because they encourage spontaneous elicitation, confrontation methods allow a finer understanding of activity, and of the individual and collective logics underlying that activity. The perspective of action is directed, rather, toward the workers. By commenting on their own

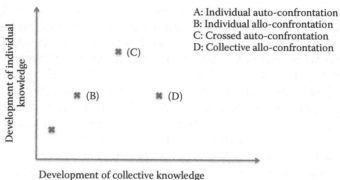

Figure 14.1 A classification of allo-confrontation methods, depending on the type of knowledge (individual or collective) that they foster.

activity, by being confronted with the activity of their colleagues, or by having to elicit their own activity to one or several colleagues, the workers gain a better understanding of their activity, alter their knowledge and construct new knowledge. In so doing, they transform their own work.

By definition, individual allo-confrontation does not allow the construction of shared knowledge. The potential for individual development is also restricted, in the sense that the worker comments on his or her own activity. However, for all the other methods we have presented, confronting workers with an activity that is not their own allows them to alter their representations and knowledge, and to construct new knowledge and representations (see Figure 14.1).

Crossed allo-confrontation is certainly the most efficient method for developing individual knowledge, in the sense that the small number of participants leaves lots of time for exchanging.

Individual allo-confrontation is the method that has the least potential for the construction of collective knowledge, as the worker is alone in commenting on the activity of a colleague.

Crossed allo-confrontation and collective confrontation may be very efficient methods for developing collective knowledge, with an added benefit in the case of the second method, in the sense that crossed allo-confrontation is limited to two participants.

Examples of applications of methods to assist reflective practice

This section presents a number of applications of confrontation methods used by ergonomists in various contexts. Their goal is to illustrate various forms of reflective practice, their specificities and their shared features, notably in terms of expected benefits. But they are, obviously, not the only examples to be used in ergonomics.

Reflecting on practice based on filmed activities

In the studies described here, reflective practice relies on the analysis of films representing the work activity. The goal is to construct shared solutions (technical, organizational, etc.) that take into account the reality of work.

The collective, video-assisted, reflective activities (ARCAVs) developed by Mhamdi (1998) are examples of this. They take place within collective meetings between workers and the hierarchy. Mhamdi's study aimed to analyze accidents with an electrical origin in a large company for the production, transportation and distribution of electricity, in order to reduce the occurrence of accidents of this kind. Films were shown during the meeting, representing interventions carried out by the workers on an everyday basis. Some of the films were produced by the workers themselves in real-world situations. A group made up of volunteers – workers, safety engineers and middle managers – met up at regular intervals to exchange about these situations. The goal is neither to judge the workers nor to reinforce the safety rules, but to perform a critical analysis of modes of operation, to discuss the applicability, usefulness and relevance of rules with respect to the real constraints of activity, and to define possible solutions for improvement, whether they be technical, procedural or organizational in nature. The author showed that accidents were rare or nonexistent at the sites where there was a regular practice of collective discussion supported by the films of collective activity.

Video was also used as a means to support confrontations during an intervention carried out within an association of saffron producers, who wished to capitalize on local knowledge and know-how to support the renewed culture of that product in their region (Mollo and Falzon, 2004). This demand led to two major difficulties. On the one hand, producers learned through trial and error in the sense that they did not have any local knowledge and it was impossible for them to literally transfer modes of culture from other countries (because of climatic conditions). On the other hand, the producers were located very far from one another, and none of their tasks were performed in co-presence.

Three kinds of confrontation were carried out:

- Individual auto-confrontations with all of the filmed producers, in order to elicit the logic underlying their activity (gestures and tools used, strategies related to the time of saffron picking, influence of early stages, etc.)

- Individual allo-confrontations with the filmed producers and with the producers who had not been filmed, in order to make visible the diversity of know-how present in the collective and to make the producers elicit it
- A collective confrontation gathering the filmed producers, located remotely from one another

Results allowed the authors to show that individual allo-confrontation formed a tool to share know-how: examining the activity of other producers allowed each producer to access part of his or her knowledge and know-how. This method, in a sense, allowed producers to compensate for the lack of a shared location. Furthermore, individual allo-confrontation constituted a tool for training and learning: confrontation with the activity of others led the producers to either strengthen their own representations and know-how or modify them, thanks to the elicitation process it provoked. Collective confrontation, finally, allowed not only setting off a process of formalizing know-how based on the analysis of work activity, but also the producers to realize the interest of sharing individual experiences, taking ownership of video as a tool for analysis and sharing, and organizing collective sessions of flowers' pruning (separation of the pistil of the flower), so that the activity might become a resource for the construction of a collective experience.

A reflection on practices based on nonnominal situations

The two studies presented here illustrate a form of reflective activity relying on processing nonnominal situations (NNSs), that is, situations that deviate from the prescriptions or raise issues of application of this prescribed work.

The first study consisted of analyzing the running of multidisciplinary consult meetings (MCMs) in oncology. These weekly meetings gather specialists from various specialties (surgery, medical oncology, radiotherapy, gynaecology, etc.) to propose therapeutic solutions in the case of NNSs, where therapeutic frameworks are difficult to apply.

Analyzing the activity that takes place in an MCM (Mollo, 2004; Mollo and Falzon, 2008) showed that, in accordance with expectations, MCMs made it possible to guarantee the reliability of decision-making, but the benefits of MCMs go beyond assisting decision-making. The collective reasoning carried out in MCMs generates a critical cross-examination of the various alternatives proposed, making it possible to collectively define the space of acceptable solutions and the space of unacceptable solutions. Thus, it makes it possible to delineate the boundaries of the local genre, within which doctors are free to choose between the available options the option that seems to best correspond to particular situations and their own expertise.

In so doing, MCMs are a tool for individual and collective learning, as professionals are led to take into account a number of new criteria elicited by their colleagues, and to integrate the rules constructed by the collective. Because similar problems had been repeatedly dealt with in MCMs, some rules of circumstantial adaptation become stable rules that ground the 'local genre', that is, the boundary of rules of adaptation that are viewed as acceptable.

The second study deals with the method of differential judgement of acceptability (DJA) proposed by Nascimento (2009). This method is one form of collective confrontation carried out based on written NNS scenarios.

Analyses were carried out in the field of radiotherapy, a specialty involving multiple professions (radiotherapists, medical physicists, dosimetrists, x-ray operators, etc.). These analyses showed that when faced with the same NNS, the judgement of different professionals relative to the acceptability of the situation diverges (Nascimento, 2009; Nascimento and Falzon, 2008) in ways that are related to their activity. Differences in judgement related to discrepancies lead to extensive discussions, which make it possible to describe the work in the real world, and its constraints, supporting the sharing of knowledge and the development of individual and collective skills. As a tool for action, via the judgement of the space of acceptable practices, the DJA allows professionals to define together the boundaries of reliability of their work system. Finally, the method supports the development of a collective: professionals take into account the possibilities and constraints of their colleagues and integrate them to define a space of possibilities.

The golden rules of reflective practice

Instituting a collective reflective practice so that it might contribute to the sustained development of individuals and organizations equates with organizing the spaces that will make it possible to debate the trade-offs that are made by workers in order to respond to the real-world conditions of performing the work activity. However, to do this, a number of conditions must be met, which we will specify here. This is not a comprehensive list, but we do believe that failure to comply with one of the rules described below will prevent the analyst from achieving the benefits described above. These golden rules do not aim to describe in detail how the methods used should be implemented in technical terms, but to specify the boundaries within which these methods may be deemed constructive.

Focusing on the real aspects of work activity

The object of reflective practices must be work activity in the real world. To avoid 'drifting' toward a general discussion about work and life in the organization, this practice may be supported by films, pictures or accounts of situations that emphasize the real conditions in which the work is carried out.

The goal may be to deal with singular situations or, alternately, with the repeated occurrence of similar situations. The crucial point is that work should be a focus of debate. However, two main kinds of situations may be subject to a debate:

- Nonnominal situations, in order to place at the centre of a debate the contradictions that workers live with while at work (Detchessahar, 2011) and to assess the acceptability of the various possible trade-offs.
- Situations that are viewed as significant by the agents, which may help capitalize the practices that work well (Gaillard, 2009)

A regular and perennial collective

For reflective practice to occur over the long term, reflective practice implies the existence of a regular and perennial collective. This condition is determined by three major factors.

First, as we have pointed out above, reflective practice must be regular in order to deal with a wide range of situations. Furthermore, this regularity makes it possible to protect the collective of work and to maintain the common frame of reference. Finally, as pointed out by Detchessahar (2011), discussions that are too far between will inevitably lead to moving the discussion away from the everyday problems of work and toward some more general information related to the life of the organization.

Second, the debate regarding work implies the symmetry of relations between the various members, even though these relations may be asymmetrical within the organization (Maggi, 2003). This condition is a requisite for ensuring the freedom of speech of agents and to avoid value judgements. Indeed, the goal is to understand the activity from the point of view of the difficulties that workers must deal with, not to reinforce rules or remind workers of their existence. This implies that when the hierarchy is present, it should adopt an attitude of understanding, not one of prescription.

Third, reflective practice suggests the voluntary participation of all the people involved, the existence of well-defined roles, and the long-term commitment of every participant. Indeed, it is important that the group be relatively stable and small to ensure proper and dynamic exchanges (Maggi, 2003), and that it will be possible to follow up on the actions that have been initiated. This does not preclude the idea of a 'variable topology' of the group, depending on the topics that are being addressed and the internal mobility of the members, but the composition of the group should be properly thought out and suited to the object of its focus.

The joint elaboration and evaluation of solutions

The interesting aspect of collective reflective activity is that it constitutes a space of confrontation between prescribed work (i.e. the task) and real work (the activity), as well as between the modes in which the activity may be carried out. In other words, its goal is to analyze variability – not to suppress discrepancies between prescribed work and real work, but to better understand them and manage them in a way that is conscious and reasoned.

However, reflective analysis must also aim to collectively develop technical solutions (e.g. purchasing equipment) or organizational solutions (allocation of tasks, schedules, training, etc.), some of which may be the object of experimentation. This involves confronting the logic of the workers with that of the other agents in the organization, in order to work on requisite adaptations.

This dual goal of analysis and action is a crucial condition for obtaining the support of the hierarchy and for taking into account real-world work in organizational evolutions. However, this does imply some involvement and some commitment on the part of the hierarchy.

The involvement and commitment of the hierarchy

To allow the sustained development of individuals and organizations, collective reflective practice must have a place in the organization, and must be encouraged and supported within the organization.

It must be a tool for the organization, one that is supported by management so that material and human resources can be devoted to it (Detchessahar, 2011) in such a way that the solutions that emerge from these reflections can be encouraged and experimented on. Giving such a status to reflective practice implies that the organization should agree to exhibit its contradictions and disruptions in order to discuss and defeat them. As Gaillard (2009) has pointed out in the case of some forms of feedback on experience, this involves 'acknowledging that the error that has been "called out" is a source of progress … and that such disruptions also exist close to home. One must be able to bear and discuss this state of matters' (our translation).

The effective involvement of managers in these spaces constitutes one of the ways in which some value can be given to the contents of these exchanges and to the solutions that are constructed at the level of upper management (Clergeau et al., 2006). In some cases, it may prove useful and even necessary to train some members of upper management to the approach of work analysis, so that they are able to host the debates based on real-world work, and to host the debates based on the reality of work, and to show the value of the evolutions constructed at the level of the organization.

Conclusion

Developing a collective reflective practice so that it might become a source of progress for organizations leads us to consider this kind of practice as a managerial activity (Gaillard, 2009; Detchessahar, 2011), to 'organize the work of organizing' (de Terssac, 2002). This implies that the knowledge that is mobilized and constructed by reflective spaces not only is useful within the company, but also may serve as a basis for the development of knowledge or tools that make it possible to transform the organization. This also implies the need to involve management in the organization of this action, so that the solutions constructed may be defended at the level of upper management and lead to concrete transformations, supported by all of the levels of the organization's hierarchy (Detchessahar, 2011; Daniellou, 2012).

The benefits derived from reflexivity and from the implementation of solutions to improve the conditions of work are not without effect on the construction of health at work. Indeed, it seems that the construction of mental well-being might be based upon the abilities that are available and can be mobilized, that is, the opposite of 'cognitive misery'. Being of sound cognitive health implies 'being competent', that is, being able to 'make use of skills that allow one to be hired, to succeed, and to make some progress' (de Montmollin, 1993, p. 40, our translation). From our point of view, being able to debate the constraints and resources of work in the real world fosters the development of skills as noted above, but beyond that, to foster the development of men and women at work, collectives and organizations. A virtuous circle is created. Because they are more competent and have access to favourable work conditions, workers are able to meet the desired performance. Because of this, they feel better because they work better.

References

Amable, B., and Askenazy, P. (2005). *Introduction à l'économie de la connaissance.* Unesco Report: Construire les Sociétés du Savoir. Paris: UNESCO.

Barthe, B., and Quéinnec, Y. (1999). Terminologie et perspectives d'analyse du travail collectif en ergonomie. *L'Année Psychologique,* 99, 663–686.

Caroly, S. (2010). L'activité collective et la réélaboration des règles: des enjeux pour la santé au travail. Habilitation thesis, Université Victor Segalen Bordeaux 2, France.

Clergeau, C., Detchessahar, M., Devigne, M., Dumond, J. P., Honoré, L., and Journé, B. (2006, November). *Transformation des organisations et santé des salariés: proposition d'un programme de recherche.* 17th congress of AGRH, Le travail au cœur de la GRH. Reims, France.

Clot, Y. (2008). *Travail et pouvoir d'agir.* Paris: PUF.

Clot, Y., Faïta, D., Fernandez, G., and Scheller, L. (2000). Entretiens en autoconfrontation croisée: une méthode en clinique de l'activité. *Pistes,* 2(1).

Daniellou, F. (2012). *Les facteurs humains et organisationnels de la sécurité industrielle: des questions pour progresser.* Cahiers de la Sécurité Industrielle 2012-03, Fondation pour une Culture de Sécurité Industrielle. Toulouse: France. Retrieved from http://www.FonCSI. org/en/cahiers/.

Daniellou, F., Simard, M., and Boissières, I. (2011). *Human and organizational factors of safety: a state of the art.* Cahiers de la Sécurité Industrielle 2011-01, Fondation pour une Culture de Sécurité Industrielle. Toulouse, France. Retrieved from http://www.FonCSI.org/en/.

de Montmollin, M. (1993, September). Compétences, charge mentale, stress: peut-on parler de santé "cognitive"? Presented at 28th Congress of SELF, Geneva, Switzerland.

de Terssac, G. (2002). *Le travail: une aventure collective.* Toulouse: Octarès.

Detchessahar, M. (2011). Santé au travail. *Revue Française de Gestion,* 5(214), 89–105.

Devulder, C., and Trey, P. (2003). *Organiser la production en équipes autonomes.* Saint-Denis-la-Plaine, France: AFNOR.

Doise, W., and Mugny, G. (1981). *Le developpement social de l'intelligence.* Paris: Inter-Editions.

Falzon, P. (1994). Les activités méta-fonctionnelles et leur assistance. *Le Travail Humain,* 57(1), 1–23.

Falzon, P. (2005, May). Developing ergonomics, developing people. Presented at 8th Southeast Asian Ergonomics Society Conference SEAES-IPS, Denpasar, Bali.

Falzon, P. (2007). Enabling safety: issues in design and continuous design. *Cognition, Technology and Work,* 10(1), 7–14.

Falzon, P., and Mollo, V. (2009). Para uma ergonomia construtiva: as condições para um trabalho capacitante. *Laboreal,* 5(1), 61–69.

Falzon, P., Sauvagnac, C., and Chatigny, C. (1996, June). Collective knowledge elaboration. Presented at Second International Conference on the Design Cooperative Systems, Juan les Pins, France.

Falzon, P., and Teiger, C. (1995). Construire l'activité. *Performances Humaines et Techniques,* Special issue, 34–40.

Gaillard, I. (2009). S'organiser pour apprendre de son expérience. In G. de Terssac, I. Boissières, and I. Gaillard (Eds.), *La sécurité en action* (pp. 151–174). Toulouse: Octarès.

Garnier, P. H. (2005). Conflit socio-cognitif et système de soin. In O. Ménard (Ed.), *Le conflit* (pp. 143–156). Paris: L'Harmattan.

George, C. (1983). *Apprendre par l'action.* Paris: PUF.

Hoc, J. M. (1996). *Supervision et contrôle de processus: la cognition en situation dynamique.* Grenoble: PUG.

Maggi, B. (2003). *De l'agir organisationnel. Un point de vue sur le travail, le bien-être, l'apprentissage.* Toulouse: Octarès.

Manin, B. (1985). Volonté générale ou délibération? Esquisse d'une délibération politique. *Le Débat,* 33, 72–93.

Mhamdi, A. (1998). Les activités de réflexion collective assistée par vidéo: un outil pour la prévention. Doctoral dissertation, CNAM, Paris.

Mollo, V. (2004). Usage des ressources, adaptation des savoirs et gestion de l'autonomie dans la décision thérapeutique. Doctoral dissertation, CNAM, Paris.

Mollo, V., and Falzon, P. (2004). Auto- and allo-confrontation as tools for reflective activities. *Applied Ergonomics*, 35(6), 531–540.

Mollo, V., and Falzon, P. (2008). The development of collective reliability: a study of therapeutic decision-making. *Theoretical Issues in Ergonomics Science*, 9(3), 223–254.

Nascimento, A. (2009). Produire la santé, produire la sécurité. Développer une culture collective de sécurité en radiothérapie. Doctoral dissertation, CNAM, Paris.

Nascimento, A., and Falzon, P. (2008, June). Reliability assessment by radiotherapy professionals. Presented at Healthcare Systems, Ergonomics and Patient Safety International Conference 2008 (HEPS 2008), Strasbourg, France.

Pastré, P. (2005). Introduction. La simulation en formation professionnelle. In P. Pastré (Ed.), *Apprendre par la simulation. De l'analyse du travail aux apprentissages professionnels* (pp. 7–13). Toulouse: Octarès.

Perrenoud, P. (2001). *Développer la pratique réflexive dans le métier d'enseignant*. Paris: ESF.

Samurçay, R., and Rabardel, P. (2004). Modèles pour l'analyse de l'activité et des compétences, propositions. In R. Samurçay and P. Pastré (Eds.), *Recherches en didactique professionnelle*. Toulouse: Octarès.

Schmidt, K. (1991). Cooperative work: a conceptual framework. In J. Rasmussen, B. Brehmer, and J. Leplat (Eds.), *Distributed decision making. Cognitive models for cooperative work* (pp. 75–100). New York: John Wiley & Sons.

Schön, D. (1983). *The reflective practitioner: how professionals think in action*. New York: Basic Books.

Teiger, C. (1993). Représentation du travail et travail de la représentation. In A. Weill-Fassina, P. Rabardel, and D. Dubois (Eds.), *Représentations pour l'action* (pp. 311–344). Toulouse: Octarès.

Teiger, C., and Laville, A. (1991). L'apprentissage de l'analyse ergonomique de travail, outil d'une formation pour l'action. *Travail et Emploi*, 47, 53–62.

Urfalino, P. (2000). La délibération et la dimension normative de la décision collective. In J. Commaille, L. Dumoulin, and C. Robert (Eds.), *La juridicisation du politique*. Paris: L.G.D.J.

Weill-Fassina, A., and Pastré, P. (2004). Les compétences professionnelles et leur développement. In P. Falzon (Ed.), *Ergonomie* (pp. 213–231). Paris: PUF.

chapter fifteen

Co-constructive analysis of work practices

Justine Arnoud and Pierre Falzon

Contents

Organizations: From a static, prescribed structure to a dynamically co-constructed system

Classical theories describing organizations tend to limit an organization to its structure, i.e. a hierarchy and a set of rules that should be applied to achieve desired goals. These theories are at the source of Taylorist organizations, and are still often present in companies to this day (Petit, 2005). In such organizations, the individual appears as a cog in the organizational 'machine'.

Representations of work organizations have evolved since the 1980s, under the joint influence of management science, sociology and ergonomics. This transformation of representations has many different sources.

On the one hand, the instability of the environment, coupled with technological transformations, has led to a greater premium being placed on organizations' ability to evolve and innovate. In this context, the human capital and processes of sustained knowledge development have emerged as elements essential to the development of organizations. The model of human capital, originally proposed by Becker (1964), posits that an organization's capital also includes the knowledge and know-how of people working therein. This capital self-generates through use. Experience allows each worker to improve his or her human capital. The organization may then choose to 'invest' in this capital, to make it 'fructify' by providing conditions that are favourable to its development. This model was reused in theories of organizational learning (Argyris and Schön, 1978) and learning organizations (Senge, 1990), all of which aim, precisely, to develop the human capital.

On the other hand, new models have strongly questioned the structural view of work organizations. Structuration theory (Giddens, 1984) and social regulation theory (Reynaud, 1989) have highlighted the importance of dialectics between organizational structure and the actions that are carried out within.

Giddens (1986) is one of the first authors to have considered structure (i.e. rules and resources) and individual actions as the 'two joint poles of the same duality'. Structure is, at the same time, the framework within which interactions take place, and the result of these interactions. In other words, 'the structural properties of social systems are both the medium and the result of the practices which they organize' (Giddens, 1984).

Similarly, the theory of social regulation proposed by Reynaud (1989) and developed by de Terssac (de Terssac and Maggi, 1996; de Terssac, 2003) ties together the structural dimension with the practices that are at work within the organization. In these models, an organization is presented as the result of constant trade-offs between explicit, official rules that emanate from prescribing agents or structures, and the rules constructed on an everyday basis by agents within the organization, which appear in reaction to prescriptions, and are dependent upon these agents' own needs for action, the events that they must face, and the lack of effectiveness of prescriptions. Thus, the organization is viewed as the product of constant social dynamics that are internal to it.

Following this view, an organization can be compared with what Rabardel and Béguin (2005; Bourmaud, this volume) call an 'instrument', whereas the structure can be viewed as an 'artifact' (e.g. models, software programs, symbols, rules, etc.) that interacts with the schemes of use that are developed by operators to cope with the work situations they encounter. The development of schemes of use in operators, as well as mechanisms of appropriation, allow each worker to tailor the artifact to his or her own needs and alter it to make better use of it.

An organizational structure may foster or hinder the development of new schemes. Sometimes, it operates in such a way that it is difficult to adapt it through one's interactions with it. Indeed, articulating the organizational structure and actions that take place therein is only possible in specific conditions. Amongst these conditions, the organization must foster learning via 'conversion factors' leading to the emergence of an 'enabling environment'. We develop these concepts below.

Toward the joint development of persons and organizations: Enabling environments

Ergonomics has a part to play in the construction of this dialectical relationship, which conveys the need to connect the 'regulated' organization (i.e. prescribed processes and procedures) and the 'managed' organization (i.e. the individual and collective actions that reorganize these prescriptions). This reorganization work is only possible if that organization provides an adequate environment to its agents, that is, if the operators can access a sufficient degree of freedom to truly make use of the resources at their disposal.

We will draw here on the work of A. Sen (2009). This author has proposed a theory of justice and freedom that is grounded in the idea of capabilities. The concept of capability relates to the entire set of operations that is truly accessible to the individual, whether he or she makes use of them or not. Thus, a capability conveys the effective possibility to make a choice. Being free implies truly having access to different options. According to Sen, the goal of public policy is the development of capabilities. Human societies should be assessed following this view.

The theory of human capital and the capabilities model exhibit strong ties with ergonomics: work activity allows the growth of skills and knowledge, and individual potential requires a favourable environment to express itself.

These models have led us to develop the concept of enabling environment as a general goal in the adaptation of work environments (Falzon, 2005, this volume; Pavageau et al., 2007). An enabling environment is defined as an environment that is nondetrimental, nonexclusive, and which allows people both to succeed in their work and to develop themselves. By contributing to the cognitive development of individuals and teams, an enabling environment encourages learning and makes it possible to broaden the capabilities of individuals, as well as their possibilities for action and choice.

Indeed, the mere existence of resources, whether they are internal to the operator (e.g. abilities, skills) or external (technical and organizational devices, colleagues, etc.), is not enough. These must be 'converted' to capabilities via specific systems: the conversion factors (Fernagu-Oudet, 2012a). These refer to 'the set of factors that facilitate (or hinder) the ability of an individual to make use of the resources that are at his/her disposal in order to convert them to concrete achievements' (Fernagu-Oudet, 2012a, p. 10, our translation). An enabling environment therefore cannot and should not define itself solely in terms of the presence of resources. It must also ensure that it is possible to convert these resources into concrete achievements.

This approach has implications in terms of methodologies for ergonomic interventions in organizational design or redesign. In Section 16.4, we will present an overall approach for such interventions and illustrate it with an example. This approach seeks to build an enabling process throughout the course of the intervention itself and, in the longer run, through the setup and development of conversion factors.

Enabling environments viewed in the light of organization studies

Using the instrumental approach (Bourmaud, this volume) to redefine organizations and the conceptual framework of enabling environments, we will argue here that an enabling environment can be defined as an environment that can be 'instrumentalized', insofar as it is open to adapt itself and can foster the emergence of an enabling mode of operation.

Following this view, the goals for ergonomics are as follows:

- First, to highlight available resources of any kind and the conversion factors at play – including those that function negatively, i.e. hinder effective and efficient use of these resources, and those which, conversely, contribute to performance, i.e. allow the mobilization of these resources.
- Second, based on this diagnosis, to implement a system to 'start up' positive and sustainable conversion factors, i.e. processes at the individual and collective levels that are liable to sustainably improve the organization.

Within this framework, the goal is not just to design environments that are suited and can be adapted to the job – but environments that are 'debatable', where the everyday 'inventions' of agents are discussed and can be integrated into the structure in such a way that design can go on in use. Therefore, the goal is to promote a 'work of organization' (de Terssac, 2003) in which the organization, viewed as an artifact, is the product of a continuous activity of rule creation, and where new rules are gradually integrated to replace old ones.

Therefore, a 'good' organization is one that can be suited to one's needs. It is an organization that can be adapted to the various situations that will have to be managed (Coutarel and Petit, 2009; Petit and Coutarel, this volume), an organization that 'agents invent every day, as much to produce a quality service, as to support their exchanges' (de Terssac, 2003, p. 133, our translation).

Supporting agents in the redesign of organizations: Co-constructive analysis of work practices

To consider an organization as something that can be 'suited to one's needs' means to infer the existence of two processes with important methodological consequences. On the one hand, operators must appropriate the artifact and revise it so as to facilitate its everyday use. On the other hand, this appropriation is potentially a source of the redesign of the organization by the operators themselves. This process of redesign may be facilitated or inhibited, depending on the possibilities offered by the structure – in particular, by the possibility of engaging in a discussion about the adjustments produced by the work of organization, using the design criteria contained within the proposed artifact. These two movements are at the root of an enabling environment. They will have to be identified and supported whenever necessary.

To achieve this, we propose a methodology that we term 'co-constructive analysis of work practices', which will be described in this section. Its goal is to implement and support a reflective practice (Mollo and Nascimento, this volume) based on observable practices. In so doing, it aims to support the gradual redesign of an organization following a developmental perspective, and a four-stage framework. The goal is first to identify existing resources – both individual and organizational – as well as the conversion factors that facilitate or prevent the effective use of these resources (stages 1 and 2). Based on this diagnosis, the goal is then to implement a system to start up positive, sustainable conversion factors (stage 3). Finally, it is necessary to observe the effects of the approach in terms of both individuals' concrete achievements and artifact modifications (stage 4).

The methodology is presented below and illustrated by an ergonomic intervention carried out within a context of organizational change (illustrations appear in text inserts). The company in which the intervention took place chose to regroup its support functions: the payroll departments of its various establishments and subsidiary companies had been gathered together in a Shared Service Centre (SSC). A SSC is a legally independent entity that carries out some or all of the tasks related to one or several of the support functions of the organization to which this entity belongs (Janssen and Joha, 2006). The goal of this change is to achieve economies of scale, while allowing operational units – who have become 'customers' – to concentrate on their core business. The intervention began just over 1 year after this change: the SSC was encountering major difficulties, and the organization was going so far as to question its viability. The human resources department, as well as the managing departments of some of the customer units, wished to understand how and why the problems were becoming so severe, and sought to improve the functioning of the new organization.

Stage 1: Initial observation of the structure and its possibilities

To start with, the goal is to identify the characteristics of the organizational structure and its consequences on the work of human operators. The organization is seen as an artifact; the features of the artifact and their impact on work activity are analyzed. Interviews with designers and decision-makers are carried out. Prescriptive documents – and more generally, any document aiming to manage work – are collected and analyzed. An analysis of work 'pre- vs. post-reorganization' is carried out. Pre-reorganization analyses can be done in the real world, in sites where it has not yet taken place. They can also be done via retrospective interviews.

THE SHARED SERVICE CENTRE ARTIFACT AND ITS EFFECTS ON ACTIVITY

Within the SSC, a preliminary diagnosis was carried out in order to understand the way in which the new organization had been implemented, and to identify its resources and potential.

According to management, the implementation of a SSC had emerged as a self-evident solution. This model had worked well in other companies, and the time had come to use it by calling on external consultancies and other specialists in these matters. Stakeholders in the organization had been subjected to this change, and it quickly led to a number of difficulties, involving both the customer units and the SSC itself.

It was then decided to clearly divide the units from the SSC. This division was enforced through signing service agreements. This implies that customer needs are known beforehand, and that the supplier agrees to deliver a service following specifications within a set time, and for a set price (Sardas, 2002). Thus, SSC operators had become service providers. They were in charge of providing this service without the possibility of setting up a proper 'service relationship' – as this relationship had been deemed useless in the model.

The possibilities of choice that operators now have access to, compared with the prior situation, were analyzed following the comparative methodology proposed by Sen (2009). Two payroll services were studied: one before its integration to the SSC, and the other within the SSC itself (Arnoud and Falzon, 2012). Results show that available options tend to be less numerous within the SSC. Technological change has altered working practices, forcing operators to process the payroll on the computer screen only, using inflexible tools. Furthermore, collaboration between the customer and the service provider, as it is organized (dividing up tasks, prohibiting phone calls), does not, on the one hand, allow operators to organize it, and on the other hand, prevents the production of a quality payroll.

The model underlying the SSC artifact has led to a strong compartmentalization of the activity of payroll managers and to a diminishing level of flexible resources – thus, to negative conversion factors.

Stage 2: Identifying the processes of redesign in use

Stage 2 consists of identifying the attempts made by operators to 'suit the organization to their needs' and to determine whether these attempts do or do not contribute to a gradual redesign of the organization. The goal here is to show, following the perspective proposed by Rabardel and Béguin (2005), how and in what ways design continues throughout use. These attempts at reorganizing are signs of the major problems experienced by the operators. They indicate 'hindrance areas', which they aim, more or less secretly, to bypass or cancel. Interviews and an ergonomic work analysis can support a data collection process focused on attempts of redesign-in-use.

The set of regulations that we observed allowed us to show that the SSC, which had been designed by other people, had been redesigned by agents in the organization so as to facilitate its everyday use. The operators carried out an appropriation of the 'SSC artifact'. According to them, a quality payroll can only be achieved with help from the customer, who is considered a partner in this activity. This appropriation plays the part of a conversion factor. Exchanges and negotiations with the customer, carried out over the phone, allow operators to increase their ability to do better. The customer is a resource, but had not necessarily been thought of in this way during the design of the organization. Therefore, discrepancies can be identified between the uses of the SSC as foreseen during design and the appropriation of the artifact in work situations. Yet, although these

THE GRADUAL REDESIGN OF THE NEW
ORGANIZATION BY ITS OPERATORS

As we were identifying a reduction in the number of available choices, operators developed new uses of the SSC structure, aiming to introduce 'enablement' through an extension of their capabilities. These attempts, in some cases, have led to the alteration of some of the prescribed principles of operation. Here is an example.

The imposed instructions have not all been upheld by the operators – in particular, the instruction aiming to prohibit telephone conversations with customers. This instruction was made more flexible, as managers quickly realized that communication with customers was a condition of success for the new organization. Telephones were not withdrawn, and operators did not hesitate to use them when needed: 'We're not supposed to talk to our customers on the phone, but we're fighting for it' (SSC operator).

regulations may have modified some of the principles of the organization, the resources of the work milieu and those of the individual remain separate. Operators wish to call customers, with the shared belief that 'we could work better together'. The organization divides up the tasks, limits the means of communication (the telephone is considered to be a waste of time) and remains convinced that the development of new tools will suppress the need for these interactions. Attempts made by operators to modify their environment so as to convert the customer-provider relationship into an opportunity are viewed by decision-makers as instances of violations of procedures and are not discussed. They remain insufficient to support a mutual understanding between the partners and to allow the construction of a 'transverse collective'. This type of collective 'relies on a work of articulation between agents, and on constant adjustments ... and implies an articulation between professions' (Motté, 2012, our translation). Here, the articulation involves customers (local correspondents) and providers (operators in SSC), who are led to do different things, but in a coordinated fashion (Lorino and Nefussi, 2007). Thus, some difficulties remain: interpersonal difficulties, discordant stories, blame attribution, etc.

In the case of the SSC, the structure, as it was initially designed, limits the possibilities of transforming available resources into effective opportunities for action and into organizational redesign in use. The wishes of operators are at odds with the few opportunities afforded by the organization. This is mostly due to the characteristics of the SSC model and of the client-provider relationship that has been implemented.

Stage 3: Co-constructive analysis of work practices

The attempts made by operators to redesign the work organization are neither easy nor always possible. Therefore, it is advisable to set up a method that will support attempts at redesign in use – not only by justifying their relevance to hierarchy, but also by supporting agents. The goal, then, is to trigger the positive and sustainable factors of conversion identified at the diagnosis stage, and notably to foster the emergence of a transverse collective.

To achieve this, a method of co-constructive analysis of work practices may be used. This method aims to start off a debate between the various professionals involved in the new organization – in this case, customers and suppliers. As for methods aiming to support reflective practices (Mollo and Nascimento, this volume), the goal of co-constructive analysis is twofold: first, to ensure an improved visibility of the work of others, and second, to allow the construction of work practices that are acceptable to all, in order to redesign the organization. This analysis is termed constructive because it fulfils the following criteria ('golden rules'; Mollo and Nascimento, this volume). It focuses on work activity, it is grounded in the will of and in the attempts made by operators, it aims to develop new organizational solutions, and it requires managerial support before it can be implemented and operated sustainably.

The approach we propose is as follows. First, visits of work sites are organized during which pairs of operators who usually interact remotely are made to meet (obviously, this requires their prior agreement). The operator being visited carries out the tasks of the day, while verbalizing his or her activity and explaining his or her constraints, the difficulties encountered, the criteria used, etc. The visiting operator observes the activity and listens to the verbalizations it gives rise to. He or she may intervene when desired to request further explanations or to provide these explanations, etc. On another day, the situation is inverted. The visited operator becomes the visiting operator, and vice versa. The ergonomist is present and collects the exchanges between the operators.

Later on, the difficulties identified and the new practices that have been considered must be debated within work groups involving the operators who took part in this system, and more generally, the teams involved, including the close hierarchy.

Co-constructive analysis of work practices combines several methods:

- A method consisting of producing verbal reports concurrently to the activity (Ericsson and Simon, 1984; Leplat and Hoc, 1981). This allows subjects to externalize the internal processes that are at work during action. It makes visible the mental activities underlying subjects' conducts. This elicitation work is useful both for the visited operator and for the visiting operator, who becomes better able to understand the activity and constraints of another.

230   Justine Arnoud and Pierre Falzon

- A reflective method similar to crossed allo-confrontation, in which each person is confronted with the activity of his or her partner (Mollo and Falzon, 2004; Mollo and Nascimento, this volume). One original aspect of the method is that reflective practice rests here on the activity of another person in real-world situations – not, as is usually done, on a recording of this activity. This joint presence allows subjects to interact with each other as the activity is carried out.
- A method of inquiry, in Argyris and Schön's (1978) sense. This method makes it possible to set as an object of discussion – and to resolve – doubts or conflicts related to the joint activity of operators. These doubts are made visible by a 'confused certainty that one could do better' (Lorino, 2009, p. 93, our translation). Recomposing, for a short time, an activity that is physically performed apart supports an inquiry and the search for solutions aiming to 'act better together'. The visiting operator is faced with his or her partner's activity and may reflect on his or her everyday doubts or may be surprised by this direct observation. The observed situation may not match with the expectations or representations of the visiting operator. Exchanges between partners may yield new ways of thinking and new actions will be discussed. Operators successively take on the role of an agent (and not of a mere spectator), actively aiming to understand the joint activity and improve it.

Many benefits can be expected from the co-constructive analysis of work practices:

- For operators, verbal reports make it possible to 'talk about work'. This supports the work of externalization, where activity is reified and made exterior to the operators (Falzon, 2005). Here, verbal reports are directed to a partner. Thus, work is made visible and can be discussed. These discussions can lead to a true work of organization (de Terssac, 2003), through the novel and shared reconstruction of procedures, rules and ways of doing things.
- For the organization, the method may lead to the creation of a shared culture, in the sense that everyone is present in each person's activity (Nascimento and Falzon, 2011). This shared culture is a guarantee of the quality and continuity of service.
- Finally, for analysts, the position chosen is quite specific. His or her contribution is more that of supporting than teaching. The analyst helps operators extend and deepen their investigations regarding the organization (Argyris and Schön, 1978). In so doing, the methodology deployed aims to involve operators in redesigning the organization. This takes place within the framework of participatory design, whose interest and effectiveness have often been demonstrated in ergonomics.

IMPLEMENTING A CO-CONSTRUCTIVE ANALYSIS BETWEEN THE CUSTOMER AND THE SUPPLIER

In the case of the shared service centres, further observations suggested evidence of symptoms of a disrupted collective activity. Numerous discrepancies were observed at the end of the line. Payroll managers did not have access to the required information in due time, or received information that was useless and ambiguous requests, etc. The concurrent observation of their customers showed that these customers did not have correct representations of the needs of the payroll managers and could not think of the activity of the managers in relation to their own. Numerous discordant stories were identified. This led not to the improvement of the joint activity, but more to the search for culprits. Each person involved was questioning the optimal character of this activity, but did not necessarily have the means to transform it. The SSC structure and its procedures often prevented the launch of 'spontaneous inquiries' between customers and providers. Yet, new practices did emerge. Payroll managers did not hesitate to use the telephone, and unofficial meetings were organized, aiming to better understand the activity of another, and the effects one's activity might have on it. The method of co-constructive analysis of work practices was imagined based on these emerging practices and on the often-repeated wish of operators to 'see each other', 'know each other' and 'understand each other'.

Instructions were given to every participant to observe the activity of his or her partner, who was requested to verbally report his or her activity. The observer could ask questions at any time. The operators invited to take part in this work all agreed to do so, and expressed their expectations: 'Tomorrow, it'll be interesting for me as well, because I will see what issues he has to deal with. Because maybe when I'm doing something, when I'm sending something, I'm thinking to myself that the fact that I'm not sending this or not telling him that, might cause problems with his payroll editing' (SSC customer).

Analysis of the dialogical activities produced during this work revealed that the participants engaged in a reflective analysis focusing on their joint collective activity (Lorino, 2009). Over the course of this work, the method allowed, on the one hand, each person to analyze his or her own activity with respect to their partners', and on the other hand, to take part in a collective reflection regarding this collective activity – aiming to assess it and, possibly, transform it.

Stage 4: Capitalizing and debating results

The final stage consists of highlighting and debating the results obtained in the previous stage. The role of the ergonomist is twofold: first, to observe the effects of coanalysis on the everyday work of the agents involved, and second, to highlight these effects in order to have managers take over with running the method. The presentation of the results is an ideal moment to bring together all of the people involved in this work and to make its legitimacy recognized within the organization. Co-constructive analysis of work practices cannot be done without operator participation. But managers should allocate the necessary time and resources to support or systematize site visits if need be (e.g. in case of difficulties in carrying out the joint activity, of discrepancies between results and expectations, or of lack of understanding). The goal of this stage is also to capitalize on the new practices resulting from these various meetings. Team meetings, where each person discusses the transformations brought about following the site visits, may help in the diffusion of good practices and in the transformation of the organization as a whole.

CONTINUATION AND DEPLOYMENT OF THE METHOD

Following the implementation of this system, new observations were carried out on customers and suppliers. This revealed the presence of a 'shared culture': the operators changed their ways of working by integrating the needs of their partner. Every operator was reassured as to how the work of his or her partner was taking place. This led to the transformation of 'discordant stories' into 'compatible stories'. These meetings allowed the elaboration of a shared vocabulary and adjustments between stakeholders. Thus, it created favourable conditions for the emergence of a transverse collective. The main results of the method were presented to decision-makers and management, and the system is operational to this day.

The method we have developed seems to have supported the activation of conversion factors. This allowed operators to make use of the resources of the transverse collective in order to convert them to opportunities. These opportunities seem to actualize themselves in the conducts and achievements of the partners involved. Today, everyone involved has access to more resources to produce quality work, as it is possible to integrate the activity of partners in one's own practice. By approving and supporting the continuation of this system, management has recognized its interest and legitimacy, both for individuals and for the organization as a whole.

The role of the ergonomic intervention in the joint development of organizations and people

Fitting the artifact to one's own activity is a means for operators to convert potential resources into capabilities and to broaden the scope of possibilities. However, this process must be taken into account by the organization. Such recognition is necessary to provide new resources to individuals and collectives, supporting the conversion of these resources into capabilities. In this way, the organization will be able to fully benefit from these capabilities and improve its operation.

Following this logic, the goal is not just to design enabling environments that are suited and adaptable, but to design environments that are debatable. Organizational changes are particular moments in the life of organizations where one can hope to be able to promote the construction of such environments (Fernagu-Oudet, 2012b). Change is favourable to learning, provided it 'consists in designing not a new organization, but a system of experimentation and learning to provoke and to foster new modes of operation in the organization' (Sardas and Lefebvre, 2005, p. 285, our translation). Yet, few organizations choose to follow this path. Often, change is imposed on operators in ways that do not support appropriation processes (Bernoux, 2004). It is precisely in such post hoc setups that ergonomists are most often called to the fore. Their goal is then to detect resources and factors that foster or inhibit the conversion of resources into effective possibilities. In other words, the aim is to carry out a reflection about the constraints and opportunities afforded by the situation (Fernagu-Oudet, 2012b). From there, an ergonomic intervention can be constructed, aiming to 'activate' conversion factors that might transform these resources into capabilities.

The goal of an ergonomic intervention aiming to (re)design an organization is twofold:

- During the implementation of the new system, the ergonomic intervention seeks to develop the capabilities of all the persons involved, through the elicitation of work, reflective practices and the joint construction of a range of possibilities.
- Following the implementation of the system, the capabilities constructed in this way express themselves in the achievements or conducts selected by the operators. For example, these capabilities may support the integration of each person's work in the practices of the collective. As a result, possibilities of 'doing a good one's work well' improve, along with overall system performance.

Thus, the development of individuals and organizations is perceived as the means and the end of the ergonomic intervention. To achieve this,

234 Justine Arnoud and Pierre Falzon

the stance of ergonomists must be threefold: an ergonomist must be a 'revealer' of resources, an 'activator' of conversion factors and a 'mediator' between the various agents in the organization.

References

Argyris, C., and Schön, D. A. (1978). *Organizational learning: a theory of action perspective.* New York: Addison-Wesley.
Arnoud, J., and Falzon, P. (2012). Shared services center and work sustainability: which contribution from ergonomics? *Work,* 41(Suppl. 1), 3914–3919.
Becker, G. (1964). *Human capital. A theoretical and empirical analysis, with special reference to education.* Chicago: University of Chicago Press.
Bernoux, P. (2004). *Sociologie du changement dans les entreprises et les organisations.* Paris: Seuil.
Coutarel, F., and Petit, J. (2009). Le réseau social dans l'intervention ergonomique: enjeux pour la conception organisationnelle. *Management et Avenir,* 7(27), 135–151.
de Terssac, G. (2003). Travail d'organisation et travail de régulation. In G. de Terssac (Ed.), *La théorie de la régulation sociale de Jean-Daniel Reynaud. Débats et prolongements* (pp. 121–134). Paris: La Découverte.
de Terssac, G., and Maggi, B. (1996). Le travail et l'approche ergonomique. In F. Daniellou (Ed.), *L'ergonomie en quête de ses principes. Débats épistémologiques* (pp. 77–102). Toulouse: Octarès.
Ericsson, K. A., and Simon, H. A. (1984). *Protocol analysis. Verbal reports as data.* Cambridge, MA: MIT Press.
Falzon, P. (2005, December). Ergonomics, knowledge development and the design of enabling environments. Presented at Humanizing Work and Work Environment Conference (HWWE 2005), Guwahati, Inde.
Fernagu-Oudet, S. (2012a). Concevoir des environnements de travail capacitants: l'exemple d'un réseau réciproque d'échanges des savoirs. *Formation-Emploi,* 119, 7–27.
Fernagu-Oudet, S. (2012b). Favoriser un environnement "capacitant" dans les organisations. In E. Bourgeois and M. Durand (Eds.), *Former pour le travail.* Paris: PUF.
Giddens, A. (1984). *The constitution of society.* Glasgow: Bell and Bain.
Janssen, M., and Joha, A. (2006). Motives for establishing shared services centers in public administrations. *International Journal of Information Management,* 26, 102–115.
Leplat, J., and Hoc, J. M. (1981). Subsequent verbalization in the study of cognitive processes. *Ergonomics,* 24(10), 743–755.
Lorino, P. (2009). Concevoir l'activité collective conjointe: l'enquête dialogique. Étude de cas sur la sécurité dans l'industrie du bâtiment. *Activités,* 6(1), 87–110. Retrieved from http://www.activites.org/v6n1/v6n1.pdf.
Lorino, P., and Nefussi, J. (2007). Tertiarisation des filières et reconstruction du sens à travers des récits collectifs. *Revue Française de Gestion,* 1(170), 75–92.
Mollo, V., and Falzon, P. (2004). Auto- and allo-confrontation as tools for reflective activities. *Applied Ergonomics,* 35(6), 531–540.
Motté, F. (2012, September). Le collectif transverse: un nouveau concept pour transformer l'activité. Presented at 47th Congress of SELF, Lyon, France.

Nascimento, A., and Falzon, P. (2011). Producing effective treatment, enhancing safety: medical physicists' strategies to ensure quality in radiotherapy. *Applied Ergonomics, 43*, 777–784.

Pavageau, P., Nascimento, A., and Falzon, P. (2007). Les risques d'exclusion dans un contexte de transformation organisationnelle. *PISTES*, 9(2). Retrieved from http://www.pistes.uqam.ca/v9n2/pdf/v9n2a6.pdf.

Petit, J. (2005). Organiser la continuité du service: intervention sur l'organisation d'une mutuelle de santé. Doctoral dissertation, Université Bordeaux 2, France.

Rabardel, P., and Béguin, P. (2005). Instrument mediated activity: from subject development to anthropocentric design. *Theoretical Issues in Ergonomics Science*, 6(5), 429–461.

Reynaud, J. D. (1989). *Les règles du jeu: l'action collective et la régulation sociale.* Paris: Armand Colin.

Sardas, J. C. (2002). Relation de partenariat et recomposition des métiers. In F. Hubault (Ed.), *La relation de service, opportunités et questions nouvelles pour l'ergonomie* (pp. 209–224). Toulouse: Octarès.

Sardas, J. C., and Levebvre, P. (2005). Théories des organisations et interventions dans les processus de changement. In J. C. Sardas and A. M. Guénette (Eds.), *Sait-on piloter le changement?* (pp. 255–289). Paris: L'Harmattan.

Sen, A. (2009). *The idea of justice.* London: Penguin Books Ltd.

Senge, P. (1990). *The fifth discipline: the art and practice of the learning organization.* New York: Doubleday.

chapter sixteen

Auto-analysis of work
A resource for the development of skills

Bénédicte Six-Touchard and Pierre Falzon

Contents

The transmission of incorporated knowledge

This chapter focuses on the acquisition of skills in particular contexts where the skills to be acquired combine gestural abilities with the collection of precise information on the objects and the tools of work. These skills are acquired gradually, throughout professional practice, in a way that is mostly tacit. Workers develop them without necessarily being conscious of the fact.

These skills are interesting to ergonomists for two reasons: (1) because they are key factors in effectiveness and quality of production and (2) because they include knowledge that is related to self-preservation.

Therefore, they are related to the two central goals of ergonomic practice: to ensure operative performance and to ensure the well-being of workers. Because of this, fostering the development of skills is a goal for ergonomists.

This goal is confronted with an obstacle: the incorporated nature of knowledge. Indeed, gestural knowledge feeds off the experience of situations, off their variability and diversity. They combine with rules that frame the use of the body, with trade-specific know-how, with pragmatic knowledge, with typical conducts, with types of reasoning, to form skills (de Montmollin, 1984). Accessing this incorporated knowledge is difficult. Observation is not enough, because of the refined nature of abilities. Collecting information via interviews with experienced workers is very haphazard; these workers are able to deploy an activity that is effective, efficient and relevant, but are not conscious of their modes of operation and the decisions underlying them. Because of this, the transmission of skills between experienced and novice workers is difficult. In order to perform this transmission, tutors must 'know what they know', in the case of both formal tuition and transmission in situations of work. This difficulty is compounded by the fact that the conditions of transmission are often sketchy: lack of training for experienced workers, lack of time for tutoring (Chassaing, 2010).

The question is therefore that of defining the methods that allow, on the one hand, the elicitation of incorporated knowledge in experienced workers, so that they might transfer this knowledge to novices, and on the other hand, the development, in novices, of an ability to analyze their own gestures so that they might learn more easily.

Following this prospect, this chapter proposes a method to train experienced workers and novices. This method is grounded in the self-analysis of work. We begin by presenting the conceptual frameworks that ground it, then the method itself, its use in two situations of occupational training, and finally, the conditions for its implementation.

Reflective practice, realization and conversion factors

The idea that human intelligence is characterized by reflection on one's own cognitive operations is not new. In 1923, Spearman referred to Plato and Aristotle to raise the possibility that having one's own thought as an object of thought is crucial in the acquisition of knowledge. Subsequently, Jean Piaget, in his books *The Grasp of Consciousness* (1976) and *Success and Understanding* (1978), developed a theory of the construction of knowledge from action, considering reflection on one's actions or realization as a necessary process in the acquisition of new knowledge. Piaget argues that action is an autonomous form of knowledge whose conceptualization relies on an internal realization. This is a deliberate and necessary process that leads to reorganizing knowledge to carry out the elaboration of experience.

Vygotsky (1986) and the various sociocognitive currents that followed highlight the crucial importance, in this process, of the interaction between one subject and another regarding tasks. He defines the 'zone of proximal development' as what subjects cannot learn on their own, but may discover with the help of another.

Both Piaget and Vygotsky point out the existence of 'unconscious concepts', or unconscious cognitive elements that are not – and cannot – necessarily be coded in language. In other words, conceptualization does not necessarily occur through words.

This issue can be addressed through the model of *capabilities* proposed by Sen (2009). Sen draws a distinction between capacities – what a person is capable of doing – and capabilities – what a person is truly able to do. Capability implies capacity, but a capacity is not enough to generate a capability. Capability requires both a capacity and some conditions (organizational, material, social, etc.) that allow the capacity to be used. If these conditions are met, the capacity can be realized, transformed and become a capability in a given situation (Zimmermann, 2011). If the conditions are not met, then it cannot do so.

The transformation of a potential resource (a capacity) into an effective resource (a capability) thus depends on *conversion factors*, 'factors related to the individual and/or the context in which that individual evolves, which facilitate (or hinder) the ability of an individual to make use of the resources that are available to him, and to convert them into concrete realizations' (Fernagu-Oudet, 2012, our translation). These factors may therefore exert a positive or a negative influence. A work situation may be said to be enabling if positive conversion factors are present, and less enabling if these factors are absent or if negative conversion factors are present. We will use the term *enabling environment* (Falzon, 2005; Falzon and Mollo, 2009; Arnoud and Falzon, this volume) when positive conversion factors are present.

The situation considered in this chapter may be analyzed using the enabling environment framework.

- Experienced workers have constructed resources (i.e. capacities) that allow them to act effectively. However, those resources are mostly inaccessible to conscience. They are therefore not in a favourable, enabling situation, as far as transmitting their knowledge is concerned. The question then becomes, how can one help them construct a representation of the concepts they need to transmit that can be put into words?

- Novice workers do not have the necessary knowledge and must acquire it. Their most crucial need is therefore to construct the capacity that they lack. The question is therefore that of the cognitive tools that might facilitate this acquisition in a sustained way – that is, in a way that outlasts the tutoring situation.

This situation can thus be described in terms of conversion factors. The goal is to design an enabling environment that will allow experienced workers to convert their incorporated knowledge into knowledge that can be put into words and transmitted, and novice workers to convert their general ability to learn into the ability to conceptualize from experience. By reusing the terminology introduced by other authors, the goal is for these novice workers to acquire productive, functional skills, as well as constructive, metafunctional skills (Delgoulet and Vidal-Gomel, this volume; Falzon, 1994).

Training for auto-analysis of work

The goal we have stated above requires developing, both in the tutors and in the apprentices, abilities for reflective analysis, allowing an examination of one's own activity. The method we propose here derives from the hypothesis that the ability to analyze one's own work is a powerful tool to support reflective analysis of this sort. Training workers for the auto-analysis of their work makes it possible to speed up the acquisition of knowledge and the development of skills by broadening the level of knowledge and control of the task and the activity (Falzon and Teiger, 2011; Rabardel and Six, 1995; Teiger, 1993; Teiger and Laville, 1991).

Methods for assisting reflective practice, whether they are individual or collective (Mollo and Falzon, 2004; Mollo and Nascimento, this volume), aim to help workers realize their own know-how and elicit the rationale that underlies this know-how. Similarly, the method proposed here for auto-analyzing work does not aim to enrich the analysis performed by ergonomists, but to develop the knowledge of work activity and skills in the worker himself, in the context of incorporated knowledge that is difficult to verbalize.

This method, which is applied in the context of occupational training, is intended for both actors of the training in a work situation: the tutor and the trainee (apprentice).

On the tutor's side, the goal is to assist the realization of incorporated knowledge and the verbalization of this knowledge. On the apprentice's side, the goal is to develop an ability for self-observation, for comparing one's activity with the activity of the tutor, and for understanding the meaning of operative words. More broadly, and beyond the situation of learning, training for work analysis – as a tool for realization and reflectivity – may provide the experienced as well as the novice worker a sustained instrument to support his or her professional development.

The stages that are described below are carried out separately with the tutor and with the novice.

Stage 1: The construction of the medium by the ergonomist

In the first stage, the ergonomist analyzes the activity of workers (tutors and apprentices). The goal is to understand the crucial dimensions of activity within the work situation and to proceed with identifying the skills that are effectively mobilized in work activity (Samurçay and Pastré, 1998). During this approach, video recordings of one or more sequences that are characteristic of the activity are produced. In the examples that we will use, where gestural activity is paramount, the video recording focuses on

- The actions carried out by the worker (gestures, movements, information collection, communication, etc.)
- The effects of these actions, that is, the transformations or successive states of the object of activity (for example, in the case of the activity of cooks, it will be the dish that is being prepared)
- The tools used to carry out these actions (use and operation)
- The work space and its preparation and organization

Performing a preliminary observation makes it possible to guide the video recording of the activity toward what will be relevant for the following stage of training for auto-analysis – and therefore to choose typical actions or problem situations. At the end of stage 1, a diagram of the understanding of the work situation is constructed, highlighting the determinants of work activity to serve as a support for training. This diagram makes it possible to represent the activity and establish relations of causality between the various components of the work situation.

Stage 2: Supporting the auto-analyses

In the second stage, tutors and apprentices are trained to analyze their own work using an exercise of auto-analysis that is carried out by each participant individually, based on viewing the video recording.

This stage comprises three steps. In the first step, the ergonomist explains the diagram of understanding (produced in stage 1 by the ergonomist) and discusses it with the worker. The analysis is therefore framed, and focused on ergonomic concepts and work analysis. In step 2, based on the viewing of the video recording, the subject is requested to describe his work 'as if you had to explain it to someone who knows nothing about it' (the doppelganger method; Clot, 2001). Finally, in step 3, still based on the viewing of the recording of activity, the description of work is systematically guided by the ergonomist.

The worker is led:

- First, to describe the operations of *execution*: The operations that are carried out using the starting materials, that allow the effective transformations of the object of action. In the case of material actions, the goal is to put into words the actions, means associated with these actions and instruments, gestures and ways of doing things.
- Second, to explain and analyze the operations of *orientation* and *control* of one's own actions: That is, the planning of action, the conditions and constraints of action, the knowledge and pragmatic knowledge that guide the realization of action, causality relations and anticipations, and also the criteria used to assess the correct outcome of the action (control information) and when the goal is reached.

Thus, the analysis that is requested on the part of workers allows them to assess the organization and logic underlying their procedures, by confronting their own actions with their initial and end states, the properties of the object and explanatory laws.

The three steps of description, elicitation and analysis are required to reach the goal of conceptualizing the activity during training sessions for auto-analysis. Indeed, merely describing the activity is not enough. The questions raised by the ergonomist must lead tutors and apprentices to question their gestural abilities and the determinants of their activity.

The goal is not just to produce a verbalization that decomposes subgoals and modes of operation. It is also to implement cognitive processing and processes related to *realization*: to identify invariants (formulations of laws, action rules), abstraction of differences and resemblances between situations, abstraction of properties that may apply to other situations (generalization), or alternately, the connection between actions and their meaning.

According to Schön (1983–1994), this questioning causes the worker to ask himself or herself not only 'What is it than can be explained in my professional knowledge?' but also 'What does my professional action teach me? What can I say about it?' The assisted analysis aims for workers to acquire 'the ability to return on their own back to what they have lived, in order to analyze their know-how and to reconstruct it at another cognitive level' (Pastré, 2005, our translation).

The exchange below is taken from a dialogue of questioning which took place during a session for training for auto-analysis of a head cook. It shows that the worker spontaneously places himself from the point of view of his own action. When viewing a film where he prepared paste for a sponge cake using a 50 L robot, the first information he provides focuses on the goals of action.

The exercise of training for auto-analysis then consists in going into the details of action goals, to derive the procedure or procedures (the how) until an explanation of the meaning of the action goal(s) is reached. In the example below, when questioned about the speed setting, the worker verbalized two other actions that were necessary for setting the speed. The ergonomist then attempts to have the worker put into words the meaning of these actions, which in this case are elicited by two goal states.

Tutor: So after that, there's turning it on and setting the speed.
Ergonomist: OK, how do you set the speed?
Tutor: I use the little crank that I'm holding in my right hand.
Ergonomist: OK, so now you've used the crank.
Tutor: So after that I'll wait for two or three turns, and after that ... there. I'm raising the vat and I'm switching to a faster speed.
Ergonomist: You've switched to the higher speed, so you've repositioned the bowl, I mean the vat?
Tutor: Yes, I've put it on the safety setting.
Ergonomist: So there you cranked it two or three notches to raise the speed. And what is it that makes you turn, or do you always crank it that much?
Tutor: Yes, it's at maximum level right now.
Ergonomist: So in general, whatever the recipe, you're going to crank it up to the max?
Tutor: Oh, no!
Ergonomist: Does it depend on the mix?
Tutor: No, it depends on what you want to do.
Ergonomist: In this case you wanted the robot to run at full speed?
Tutor: That was to raise my eggs and sugar.
Ergonomist: So it needed to be very high speed?
Tutor: Yes, it needs to be high speed in this case.
Ergonomist: From the start?
Tutor: Yes, from the start. I mean, we let it run for two or three turns at least, so that the eggs and sugar have time to mix, and then we raise the speed.

Some examples of applications in occupational training

The prospects we present below were derived from two interventions in sandwich courses for workers in seawater therapy and for cooks, within their respective companies. Cooks in traditional or company restaurants produce the dishes that are served to the customers in the room. They master some basic recipes, the principles of cooking and conserving food, to produce dishes in quantities that are greater than in the case of the everyday 'housewife' (e.g. preparing 50 L of mashed potatoes). Workers in a seawater therapy centre will provide bodily care by using elements from the marine environment: seawater, muds and seaweed. There are various types of hydrotherapy care: sprinkling of water jets, baths, coating in seaweed or mud. These actions are performed based on a medical prescription, and require the acquisition of techniques for manual care whose goal is to drain, relax and tone the body of a patient or customer.

The analyses dealt with eight situations of tutoring in kitchens and five situations in seawater therapy. These situations involved junior workers who had newly joined the company or, alternately, employees in training.

We will present below the results of the auto-analysis exercise during training with tutors and apprentices, and how the results of auto-analysis were subsequently used by the tutors and apprentices.

During training sessions for auto-analysis

The exercise of auto-analyzing one's work with the assistance of the ergonomist allows two means to realize skills: either through the sole observation of activity without verbalization (self-observation) or through elicitation.

- *Becoming aware of one's own skills through self-observation.* Watching a film, without any verbalizations, allows workers to become aware of their skills in several ways.

    First, they notice errors or incidents. Thus, for example, a tutor who is head cook realizes, from the sole observation of his activity, that he has not chosen the proper knife for the task he wanted to carry out.

    Second, the film allows professional practice to be seen as an object of analysis and as a means to support knowledge. The following extract is an interaction between the head cook (tutor) and the ergonomist. It concerns the action of dicing up bacon:

    > Because ... there's my hand, in relation to the chopping board.... In the end, I'm doing it wrong. Supposing this was to teach a junior worker, it's

wrong.... Ah yes, here I had to spend more time
than usual. If I'd had a knife with a proper heel.
Because in the end, I'm working on the edge of the
board. If I explain that to junior guys and they do
it in their vocational exam, they get it all wrong. I
work with the knife raising my hand more or less.

During the training session for auto-analysis, this tutor realizes
the formative impact of his actions: he realizes that it is not enough to
show the learner how to do things. He realizes that his role is not to
provide the apprentice with models of gestures or actions to repro-
duce, or to formulate propositions that the learner should repeat
and learn. His role is that of a mediator (Vergnaud, 1992), since the
decisive role is that played by the learner. Skills should therefore be
exercised as a tool to help apprentices build up their skills.

This latter point was confirmed by the apprentices who, at the
end of the training program, stated that they had changed the way
in which they observed the activity of the tutors – this after they
had realized (through auto-observation) what one could observe in
a work situation: 'It helps, because it's true, we watch carefully, but
we don't always know what to look at'. This exercise therefore allows
them to view the situation differently, to observe the work situation
in a different way.

- *Becoming aware of one's own skills through elicitation.* Thanks to the
questions posed by the ergonomist, the tutors and apprentices discovered the possibilities of verbalizing various aspects of activity and
skills. Thus, the seawater therapy agents discovered that they collected
various kinds of sensory information. In particular, they realized the
importance of touch when performing care actions using a water jet
on a patient immersed in a tub filled with murky seawater. Because
draining the blood network cannot rely on visual information (since
the seawater is murky), the worker will grasp the jet in the right hand
in a way that allows him or her to evaluate, by either touching or
brushing past, the distance between the jet and the patient's body.

Through elicitation, the pragmatic knowledge that guides
actions is made visible, either through verbalizations that focus
on the variability of modes of operation (the field of possibilities)
or through the search for meaning. Thus, in seawater therapy, the
agents also elicit the use of their left hand when they are draining a
patient. This hand does not hold the jet but is used, above the right
hand, to be able to follow the patient's blood network when the seawater is murky.

Beyond highlighting competence, the exercise of auto-analysis
makes it possible to strengthen the need to tell and elicit elements of

one's activity (for the tutor) or to question them (for the apprentice taking part in a tutoring situation), and not to become 'locked up in the action'. Thus, a tutor noted, at the end of the exercise: 'We tend to think that there are lots of things we do through force of habit, but in fact, habits don't just turn up…. It questions the habits we have behind each gesture, which in fact we don't do out of habit. This is just because some things have been formed in the mind. That is, we practice our work this way and that, for this and that reason. From there, we might ask ourselves whether there might be another way of doing things that would be more practical or more comfortable'.

Each viewing of the film, each retrospection on their activity with the ergonomist, is an opportunity for workers to transform their point of view and to understand differently the organization of their actions (Six-Touchard, 1999).

Later use of auto-analysis in tutoring interactions in the workplace

Later effects of auto-analysis by the participants in the training session (tutors and apprentices) were identified in the development of formative interactions in on-the-job situations of transmission. These effects were highlighted by the comparative analysis of two video recordings of inter-actions in the training sessions, both in the kitchen and in the seawater therapy facility. Interactions were recorded before the training in auto-analysis of work took place, and the other took place afterwards.

Two elements transforming conditions of learning can be directly attributed to workers taking ownership of the tool of auto-analysis.

One element is the enrichment of the contents of interactions between the tutor and the apprentice. Following training to auto-analysis, trans-mission is strengthened from the didactic point of view, through a greater decomposition of gestures by the tutor and through greater precision regarding the gestures and evaluation criteria that are required to guide the apprentice in carrying out the task. This quantitative enrichment of interactions is doubled over by a qualitative enrichment. Most tutors ver-balize more rules of action when involved in a situation of transmission.

The second element of transformation that was observed is the rein-forcement of the interrogative form of exchanges, encouraging a reflection on the actions at hand. The increase in the number of questions posed by the tutor to the apprentice (and vice versa) is significant, and one can also note the appearance and development of sequences of questions and answers that are similar to those developed in the auto-analysis train-ing. For example, after having uttered the action 'You add 6 L of beer to the dough', the tutor asked the apprentice: 'Aren't you going to ask me why you need to add some beer to the dough?' A chain of questions and answers followed, allowing the apprentice to discover the meaning of that action for himself.

Following training, auto-analysis had become, for each of the participants to the training program in work situations, a resource to be subsequently used to develop their interactions. Having access to a shared tool for analysis allowed them to focus their formative interactions on objects that are shared and understood in the same ways. Therefore, the participants have access to the elements of a common language that contribute to organize and facilitate formative interaction.

Required mediations for reflective practice based on auto-analysis

Hence, the reflective approach leads workers to learn how to see, how to identify and how to verbalize the elements of their activity, and then to reflect on what they did and on how they did it. This approach relies on a triple mediation, based on the film of the activity, on the diagram of the understanding of activity, and on the dialogue and questioning. These three mediations allow workers to take a step back and conceptualize their activity, and to develop their own constructive activity. These mediations closely resemble the conditions described by Mollo and Nascimento (this volume) regarding collective reflective activities.

Preliminary analysis of the activity and construction of the media for reflection

The role of the ergonomist to support the analysis of action by workers themselves necessitates a good knowledge of work in the field that the trainee (tutor, apprentice, worker) is involved in. This knowledge is acquired through a preliminary analysis of activity, before any auto-analysis training takes place. A diagram of the activity is constructed after the preliminary analysis in order to serve as a medium for training. This diagram is displayed at the beginning of the training exercise, and allows workers to better understand what will be expected of them during the viewing of their activity. Their verbalizations are focused on work activity, thus inhibiting whatever apprehension they may have about the judgement of themselves or of their activity. Then, during the analysis, the diagram is used in two ways: the filmed reality is connected to the diagram, and the diagram makes it possible to establish relations between the various elements of the work situation.

Distancing with the reality of the work activity

The point of the exercise of auto-analysis assisted by a video recording lies in the possibility for workers to perform a confrontation between observations (given by the image) and verbalizations (provoked by the analysis) of their own work activity. The video image supports distancing oneself in relation to one's own work situation. This temporary split allows subjects to place themselves in the position of both the observer and the analyst of their own actions, in order to self-diagnose what they are doing and how, and to collect information for themselves.

Other tools might be used for auto-analysis, which would introduce different types of distancing relations, such as

- The intervention of a third party with a mutual questioning. This kind of situation is similar to the crossed auto-confrontations introduced by Clot et al. (2000) or to allo-confrontation described by Mollo and Falzon (2004). In the context of occupational training, they may also contribute to strengthening the construction of a shared language between the tutor and the apprentice.
- The introduction of additional written media for training, where the frequency of training would be suited to the skill level of each participant. Indeed, depending on the level of expertise, a single session to take ownership of reflective practice may not be enough.

Guiding the questioning toward the development of reflectivity

The dialogue and questioning by the ergonomist aims to help trainees to elicit and organize their practices from the conceptual point of view – switching from what can be done to what can be uttered, through the mediation of what can be seen, and putting their activity into words (Schön, 1983). The second goal is to support a permanent use of this approach, including in the absence of the ergonomist (generalizing the training to other tasks). This might be described as the permanent use of a positive conversion factor.

The ergonomist carries out a critical (positive and negative) analysis of the activity. Workers are encouraged to be surprised by their own activity. It is the confrontation of operators with a point of view that differs from their own that leads them to broaden, complete and modify their points of view. They discover the complexity of their activity and learn how to see it in a different light.

This exercise of individual confrontation aims, for the most part, to provide some input to the constructive activity of subjects, by providing some knowledge or structures of thought that will facilitate working with knowledge. This reflective, creative activity on one's own work is defined by Falzon (1994) as a metafunctional activity – that is, an activity 'that is not geared directly toward immediate production, but an activity focusing on the construction of knowledge and tools (both material and cognitive) that are destined for eventual future use'. By questioning workers, the ergonomist opens a path for the development of this meta-cognitive knowledge. This allows subjects not only to be productive, but also to protect themselves and develop their abilities and empower themselves. Auto-analysis of work thus emerges as a conversion factor that can be used in the long term, and training for auto-analysis, as a means for a constructive approach to ergonomics.

References

Chassaing, K. (2010). Les gestuelles à l'épreuve de l'organisation du travail: du contexte de l'industrie automobile à celui du génie civil. *Le Travail Humain,* 73(2), 163–192.

Clot, Y. (2001). Méthodologie en clinique de l'activité: l'exemple du sosie. In M. Santiago-Delefosse and G. Rouan (Eds.), *Les méthodes qualitatives en psychologie.* Paris: Dunod.

Clot, Y., Faïta, D., Fernandez, G., and Scheller, L. (2000). Entretiens en auto-confrontation croisée: une méthode en clinique de l'activité. *Pistes,* 2(1).

de Montmollin, M. (1984). *L'intelligence de la tâche, elements d'ergonomie cognitive.* Berne: Peter Lang.

Falzon, P. (1994). Les activités méta-fonctionnelles et leur assistance. *Le Travail Humain,* 57(1), 1–23.

Falzon, P. (2005, December). Ergonomics, knowledge development and the design of enabling environments. Presented at Humanizing Work and Work Environment Conference (HWWE 2005), Guwahati, India.

Falzon, P., and Mollo, V. (2009). Para uma ergonomia construtiva: as condições para um trabalho capacitante. *Laboreal,* V(1), 61–69.

Falzon, P., and Teiger, C. (2011). Ergonomie, formation et transformation du travail. In P. Caspar and P. Carré (Eds.), *Traité des sciences et techniques de la formation* (pp. 143–159). Paris: Dunod.

Fernagu-Oudet, S. (2012). Concevoir des environnements de travail capacitants comme espace de développement professionnel: le cas du réseau réciproque d'échanges des savoirs à La Poste. *Formation-Emploi,* 119, 7–27.

Mollo, V., and Falzon, P. (2004). Auto-and allo-confrontation as tools for reflective activities. *Applied Ergonomics,* 35(6), 531–540.

Pastré, P. (2005). Introduction. La simulation en formation professionnelle. In P. Pastré (Ed.), *Apprendre par la simulation. De l'analyse du travail aux apprentissages professionnels.* Toulouse: Octarès.

Piaget, J. (1976). *The grasp of consciousness: action and concept in the young child.* Cambridge, MA: Harvard University Press.

Piaget, J. (1978). *Success and understanding.* Cambridge, MA: Harvard University Press.

Rabardel, P., and Six, B. (1995). Outiller les acteurs de la formation pour le développement des compétences au travail. *Education Permanente,* 100, 33–43.

Samurçay, R., and Pastré, P. (1998, February). *L'ergonomie et la didactique, l'émergence d'un nouveau champ de recherche: la didactique professionnelle. Journées de recherche et ergonomie*. Université Toulouse le Mirail, France.

Schön, D. A. (1983). *The reflective practitioner – how professionals think in action*. London: Temple Smith.

Sen, A. (2009). *The idea of justice*. London: Penguin Books Ltd.

Six-Touchard, B. (1999). L'auto-analyse du travail: un outil de prise de conscience des compétences pour la transformation des conditions d'apprentissage. Doctoral dissertation, Ecole Pratique des Hautes Etudes, Paris.

Spearman, C. (1923). *The nature of "intelligence" and the principles of cognition*. Oxford: Macmillan.

Teiger, C. (1993). Représentation du travail – travail de la représentation. In A. Weil-Fassina, D. Dubois, and P. Rabardel (Eds.), *Représentations pour l'action* (pp. 311–344). Toulouse: Octarès.

Teiger, C., and Laville, A. (1991). L'apprentissage de l'analyse ergonomique du travail, outil d'une formation pour l'action. *Travail et Emploi*, 47.

Vergnaud, P. (1992). Qu'est-ce que la didactique? En quoi peut-elle intéresser la formation des adultes peu qualifiés? *Education Permanente*, 111, 19–31.

Vygotsky, L. S. (1986). *Thought and language*. Cambridge, MA: MIT Press.

Zimmermann, B. (2011). *Ce que travailler veut dire. Sociologie des capacités et des parcours professionnels*. Paris: Economica.

Index

Milton Keynes UK
Ingram Content Group UK Ltd.
UKHW040446071024
449327UK00020B/1030

9 780367 378325